SMALLER
FASTER
LIGHTER
DENSER
CHEAPER

SMALLER
FASTER
LIGHTER
DENSER
CHEAPER

How Innovation Keeps
Proving the Catastrophists Wrong

Robert Bryce

PublicAffairs
New York

PublicAffairs books are available at special discounts for bulk purchases in the US by corporations, institutions, and other organizations. For more information, please contact the Special Markets Department at the Perseus Books Group, 2300 Chestnut Street, Suite 200, Philadelphia, PA 19103, call (800) 810-4145, ext. 5000, or e-mail special.markets@perseusbooks.com.

Book design by Jack Lenzo

Library of Congress Cataloging-in-Publication Data
Bryce, Robert.
  Smaller faster lighter denser cheaper : how innovation keeps proving the catastrophists wrong / Robert Bryce.
      pages cm
  Includes bibliographical references and index.
  ISBN 978-1-61039-205-1 (hardback)—ISBN 978-1-61039-206-8 (e-book)
1. Technological innovations—Popular works. I. Title.
  T173.8.B76 2014
  338'.064—dc23

                                        2013049381

ISBN 978-1-161039-525-0 (international PB)

First Edition

10 9 8 7 6 5 4 3 2 1

For my mother,
Ann Mahoney Bryce

CONTENTS

PART II Our Attosecond World: How We Got Here, Where We're Going, and the Companies Leading the Way

PART III The Need for Cheaper Energy

LIST OF GRAPHICS,
TABLES, AND PHOTOS

GRAPHICS

TABLES

PHOTOS

AUTHOR'S NOTE

I like Austin Kleon's 2012 book *Steal Like an Artist: Ten Things Nobody Told You About Being Creative*. One of the lines from it resonated with me: "Write the book you want to read."

I did that here.

Kleon's book is quirky, and the one you are holding is, too. My aim was to make this book inviting and easy to read. That's why I've included so many graphics and photographs. I wanted to provide lots of entry points so that even if readers don't capture every word, they can still grasp the key arguments and understand why I'm optimistic about the future and why they should be, too.

Before I go further, a note about vocabulary. The word "density" usually refers to mass per unit of volume. Here I'm using a broader interpretation of density, so that it includes population density, agricultural density, and other metrics. Given how critical density is to our culture, we need a broader definition of "dense."

One other note about the content: where possible, I've included metric conversions so that readers from outside the United States, as well as those living here, can have the units being discussed in SI form. (SI is an abbreviation for the System of International Units.) I've also included a list of SI numerical designations in Appendix A, as Americans need to get more familiar with the nomenclature.

Now for some acknowledgments. Books, at least in my case, are solo projects. While this was a solo writing effort, it required lots of people to make it happen. As such, I have many people to thank and acknowledge. The people at the Manhattan Institute for Policy Research were wonderful. I joined the think tank in 2010 at about the same time that my last book, *Power Hungry*, was published. The affiliation has been stimulating and productive. I'm bored by the Left-Right, Democratic-Republican, liberal-conservative divide. I want to be with smart people who are promoting economic growth and liberty. Manhattan Institute is packed with smart people who are doing just that. In particular, I must acknowledge Howard Husock, MI's director of research. Howard has repeatedly shown his ability to distill complex arguments into their essential points. My other colleagues at Manhattan Institute, including Larry Mone, Vanessa Mendoza, Michael Allegretti, Matt Olsen, and Bobby Sherwood, were also extremely supportive.

The entire crew at PublicAffairs were, as usual, wonderful. They are all pros. I have been extraordinarily lucky in my book publishing career to have had a single publisher (PublicAffairs) and a single editor. I'm proud to call Lisa Kaufman my editor and my friend. Lisa has a genius for being able to read a 90,000-word manuscript, digest the entire thing, and then explain how it needs to be organized to make it better. She's the best. My other friends at PublicAffairs—Clive Priddle, Susan Weinberg (who's now the group publisher for Basic Books, Nation Books, and PublicAffairs), Peter Osnos, Melissa Raymond, Tessa Shanks, and Jaime Leifer—were also great. In addition, Collin Tracy did a great job managing the production of the book, and copy editor Jerold Kappes was thorough and patient.

I've also been lucky to have the same person doing the fact checking on all five of my books. My pal Mimi Bardagjy worked through about a thousand footnotes. She treated each one punctiliously. Better still, she kept her good humor throughout.

I've had plenty of research help. Grant Huber provided helpful data. My friend Leslie McLain was, once again, invaluable. Yevginy Feyman at the Manhattan Institute was great at providing research and graphics. George Voorhes of Red Barn Muse Creative Group in Portland made the majority of the graphics. I recommend his work without reservation.

While I had plenty of help putting this book together, any errors are mine and mine alone. If you spot a mistake, please let me know so it can be corrected for the paperback edition.

My appreciation also goes to my friend Buddy Kleemeier, who was instrumental in arranging my visit to a drill rig. Hans Helmerich and Rob Stauder were patient tutors regarding drilling-rig technology. Cal Cooper offered valuable perspective on the history of drilling and the ongoing progress being made in that sector. My friends Hill Abell and Frank Kurzawa never tired of talking about bikes and watts. Jan Van der Spiegel at the University of Pennsylvania went out of his way to send me a photo of ENIAC-on-a-chip that he and his students developed about two decades ago. John Fannin and Michael Ramos were helpful in discussing music technology and recording. I must also thank my pal and Web guru Tyson Culver, who has been instrumental in keeping me current in the digital age.

I also want to thank Joe Bruno, Mark Ehsani, Anthony Holm, Rob Manzer, Eric Topol, Anas Alhajji, and Jesse Ausubel. Others who need to be acknowledged and thanked include my longtime friend Robert Elder Jr., who patiently read many different drafts and offered encouragement and insights. Omar Kader, the CEO of Pal-Tech, also made time in his busy schedule to read over a draft of the manuscript. Stan Jakuba, who was a pivotal reviewer of the early drafts of my last book, *Power Hungry*, was also a sharp-eyed reader. So, too, was Rex Rivolo. Rex has been a friend for many years, and he offered some key technical guidance as I thought about power density. Another friend, Bruce Hamilton, provided guidance on nuclear technology and helped me avoid several errors.

In addition, my Tulsa connections—Bryan Shahan, Violet and Ronald Cauthon, Chris Cauthon, and R. Dobie Langenkamp—have always been supportive and helpful. I must also acknowledge my father-in-law, Paul Rasmussen, a professor emeritus in chemistry at University of Michigan. Even in his 70s, Paul remains one of the hardest-working people I know. He read numerous chapters and untold drafts with good humor. He was particularly helpful when it came to understanding battery technology.

I must also acknowledge my agent, Dan Green. We have been friends since 2001, when we were introduced by our mutual friend, Lou Dubose. I am proud to work with Dan. He's a pro.

Finally, I must thank my wife, Lorin, and our three children, Mary, Michael, and Jacob. Lorin and I have been married for nearly three decades. Every day I am amazed and humbled by her love and support. As for my children, no father has ever been as proud.

We are lucky to be living in extraordinary times. And because of the inexorable trend of Smaller Faster Lighter Denser Cheaper, those times are only going to become more extraordinary.

11 December 2013
Austin, Texas

MOVING BEYOND "COLLAPSE ANXIETY"

We are besieged by bad news.

Climate change, pollution, famine, water shortages, war and terrorism, the mess at Fukushima, political gridlock, and the ongoing debt problems and economic malaise in Europe and the United States are dominating the headlines. On October 31, 2011, demographers at the United Nations announced that the Earth now hosts some seven billion people, prompting UN Secretary-General Ban Ki-Moon to declare that "alarm bells are ringing."[1]

Those alarm bells are also continually ringing about the danger of pandemics and epidemics. In 2007, the head of the World Health Organization warned that new diseases are "emerging at the historically unprecedented rate of one per year," and given the ease of international air travel, she went on to say that it would be "extremely naïve and complacent" to assume that the world will not be hit by another disease like AIDS, the Ebola virus, or severe acute respiratory syndrome (SARS).[2] In 2013, two new respiratory viruses came to light—including a coronavirus in the Middle East that is similar to a bat virus, and a new strain of bird flu in China, known as H7N9—and the WHO quickly warned health officials to monitor any unusual cases of respiratory problems. Those outbreaks came on the heels of outbreaks of swine flu and a strain known as H1N1.[3]

Television news inundates us with the latest images of floods in Europe, hurricanes in New York, wildfires in Australia and the American West, earthquakes in Haiti and Japan, and drought in California and

Texas. Terrorism, or even the hint of a terrorist attack, always makes the news. The US government continually ranks the risk of terrorism with a color-coded system. In July 2013, the terror-alert chart was yellow, for "Elevated: Significant Risk of Terrorist Attacks." Terror-alert.com will even send you an e-mail whenever the alert status changes.[4] To all of those worries, add in gun violence, train derailments, fertilizer-plant explosions, the never-ending violence in the Middle East and Africa, and it seems like the drumbeat of bad news will never end.

The avalanche of bad news has led many people to experience, or even embrace, what author Gregg Easterbrook calls "collapse anxiety." Easterbrook defines the condition as a "widespread feeling that the prosperity of the United States and the European Union cannot really be enjoyed because the Western lifestyle may crash owing to economic breakdown, environmental damage, resource exhaustion . . . or some other imposed calamity."[5]

Collapse anxiety pervades the rhetoric of many of the world's most prominent environmentalists as well as some of the biggest environmental groups. They abhor modern energy sources as despoilers of earth's beauty and natural order and cling to the idea that we humans have inappropriately sought to subdue nature for our own shortsighted, materialistic, and short-term benefit. In their view, we humans have sinned so much against Mother Earth that even the weather has turned against us. Drought, wildfires, hurricanes, tornadoes are all increasing in frequency and intensity, we are told, due to climate change caused by the amount of human-produced carbon dioxide in the atmosphere. And those carbon dioxide emissions are due to the fact that we humans are using too much energy.[6] We are driving too much, flying too much, eating too much, making too much unneeded stuff, and using far too much air-conditioning and refrigeration.

The fundamental outlook behind collapse anxiety is one of scarcity and shortage. It's a view first put forward by the English economist Thomas Malthus, who forecast a dire future in "An Essay on the Principle of Population," which was published in 1798. Malthus claimed that increasing global population would soon result in starvation for many people as the world would not be able to feed itself.[7] Today's neo-Malthusians, a group that includes John Holdren, President Barack Obama's top science adviser, advocate radical approaches to forestalling catastrophe, including what they call "de-growth."[8] This worldview is frequently represented in the pages of *The Nation, Mother Jones,* and other Left-leaning media outlets.[9] It can also be seen with depressing regularity on the Op-Ed pages of the *New York Times.*[10] And it is most obvious in the prescriptions put forward by some of the world's biggest environmental groups, including the Sierra Club and Greenpeace. The worldview of the degrowthers was neatly summarized in a 2013 segment of Bill Moyers's TV show, *Moyers & Company.* It was called "Saving the Earth from Ourselves."

The prescriptions put forward by the degrowth crowd are familiar. Nuclear energy is bad. Genetically modified foods are bad. Coal isn't just bad, it's awful. Oil is bad. Natural gas—and the process often used to produce it, hydraulic fracturing—is bad. Those things must be replaced by what the degrowth crowd claims are the Earth-friendly ones. Renewable energy, of course, is good. Organic food is good. Locally grown organic food is even better. And if you really care about Mother Earth, then you will give up flying. Less air travel means less jet fuel gets burned and therefore less carbon dioxide is produced.

The mantra of the neo-Malthusians is "peak everything." In fact, a book carrying that very title, *Peak Everything: Waking Up to the Century of Declines,* by Richard Heinberg, was published in 2007. In this neo-Malthusian view, there are simply too many of us humans, and we are using too much of everything. We should—as the segment on Moyers's show put it—be saving the Earth from us. The catastrophists claim that we are running out of essential commodities—food, oil, copper, iron ore. Given our myriad sins against the planet, we are surely going to pay. This dystopian outlook appeals to plenty of people. It seems they cannot be happy unless they are scared out of their minds.

This pessimistic worldview ignores an undeniable truth: more people are living longer, healthier, freer, more peaceful, lives than at any time in human history. Amidst all of the hand wringing over climate change, genetically modified foods, the latest Miley Cyrus video, and other alleged harbingers of our decline as a species, the plain reality is that things are getting better, a lot better, for tens of millions of people all around the world.

Dozens of factors can be cited for the improving conditions of humankind. But the simplest explanation is that innovation is allowing us to do more with less. We are continually making things and processes Smaller Faster Lighter Denser Cheaper. Our desire to do more work and exchange more information is making our computers Smaller Faster. From food packaging to running shoes, nearly everything we use is getting Lighter. More precise machinery is making our engines and farms Denser. And always—always—innovators are driving down costs and making goods and services Cheaper.

The innovation that drives the push for Smaller Faster Lighter Denser Cheaper is making us richer and that, in turn, is helping us protect the environment. Density is green. And thanks to our ability to wring more energy and more food from smaller pieces of land, we can save wild places and wild things from development.

The trend toward Smaller Faster is not dependent on a single country, company, or technology. Nor is it dependent on ideology. Smaller Faster Lighter Denser Cheaper has flourished despite Marxism, Communism, Socialism, Confucianism, and authoritarian dictatorships. It might even survive the Republicans and the Democrats.

The centuries-long trend toward Smaller Faster Lighter Denser Cheaper will continue. It may even accelerate in the years ahead thanks to ever-cheaper computing, high-speed Internet connectivity, wireless communications, 3-D printing, and other technologies that are catalyzing yet more innovation.

This book is a celebration of the trend toward Smaller Faster Lighter Denser Cheaper. It's also a rejoinder to the doomsayers, a rebuttal to the catastrophists who insist that disaster lurks just around the corner. Big environmental groups like Greenpeace, Sierra Club, Natural Resources Defense Council, and others raise hundreds of millions of dollars every year by instilling fear and proclaiming that we humans are headed for disaster. Those groups and their many supporters have the right intentions—the desire to preserve nature, wild places, and rare animals—but in many cases, their proposed solutions will only exacerbate the problems they claim to be addressing.

Do we face challenges? Of course. We face a panoply of scary problems ranging from rogue asteroids and climate change to the loss of privacy in our networked age and all-out cyberwar.[11] Shortages of freshwater, excessive use of pesticides, destruction of the rain forests, and the problem of declining topsoil only add to the list of worries that can cause collapse anxiety. The bad-news list goes on and on, and the mainstream media adds to that list every day. Bad news sells. If it bleeds, it leads. No politician ever got elected by telling voters that everything is going to be just fine the way it is.

There's no doubt that we have many problems. But our future doesn't lie in the past. We cannot solve our problems by forgoing modern energy sources and eschewing modern agriculture for a "simpler life" based on renewable energy and organic food. For millennia, we humans subsisted on the ragged edge of starvation by relying on those sources. If we want to continue bringing people out of poverty, we must embrace innovation, not reject it. We need an ethic that embraces both humanism and environmental protection. We need an ethic that embraces innovation and optimism. In short, we need to embrace the ingenuity and entrepreneurial spirit that is continually making things Smaller Faster Lighter Denser Cheaper.

Examples of that ingenuity abound. The smart phone I carry in my pocket has 16 gigabytes (16 billion bytes) of data-storage capacity. That's about 250,000 times more capacity than that of the Apollo Guidance Computer onboard *Apollo 11*, the spaceship that Neil Armstrong and Buzz Aldrin used when they landed at the Sea of Tranquility on July 20, 1969, when I was nine years and one day old.[12] We are living in a world equipped with physical-science capabilities that stagger the imagination—from nanoparticle medicines that battle cancer to intra-solar-system exploration feats like NASA's Curiosity Rover, a plutonium-fueled six-wheel-drive robot that's gallivanting across the surface of Mars with as much ease as if the Red Planet were only a tad more remote than Candelaria, Texas.[13] Sequencing the human genome, which can help doctors diagnose and treat illness, has become almost routine as the process has gotten Faster Cheaper. Over the past decade or so, the cost to sequence a human genome has dropped from millions of dollars to less than $10,000.[14]

The purpose of this book is to put a name to what's happening, and to illuminate how the extraordinary discoveries and developments transforming everything from computers and cars to medicine and sports are rooted in the push for Smaller Faster Lighter Denser Cheaper.

This book provides a lens to examine and make sense of our history and our future. It showcases the innovations, individuals, and companies that are allowing us to do more with less. It lauds the tycoons of the Industrial Age and the twenty- and thirty-something inventors of today who are trying to develop and market The Next Big Smaller-Faster-Lighter-Denser-Cheaper Thing.

Yes, I am optimistic about the future. Absolutely. But I'm no Doctor Pangloss. I'm not claiming that technology will solve all our ills. It won't, and can't, force humans to love one another or, heck, even to be polite while standing in a queue. Innovation created penicillin. It also gave us the AK-47. I am leery of what my fellow PublicAffairs author Evgeny Morozov rightly calls "solutionism," the belief that all of our ills can be solved if only we have the right technology, whether that be smart phones, or algorithms, or big data sets. In his 2013 book, *To Save Everything Click Here,* Morozov writes that over the last century "virtually every generation has felt like it was on the edge of a technological revolution."[15] And over the past few years, bookstores—remember them?—have been flooded with chock-full-of-optimism tomes, from *Dow 30,000* to *Infinite Progress.*

My bias is not that we are on the edge of a technological revolution—although that may well happen—but rather that we must recognize the countless Smaller Faster Lighter Denser Cheaper technologies that have come before us as well as those that lie ahead. Improved medicines are allowing us to live longer. Faster Lighter more powerful, more efficient automobiles and airplanes are allowing us to travel farther, safer, in greater comfort. Cheap, or even free, communications technologies like e-mail and Skype are giving us the ability to communicate with nearly anyone on the planet instantaneously. We humans were born to network, and our increasing ability to network with people who are across town or a dozen time zones away, combined with cheap (or even free) computing power, is fostering countless new technologies.

The Internet is freeing information like never before, freeing men, and even more, women and girls, from the intellectual and societal chains that for centuries have been wielded by the kings, generals, priests, rabbis, and mullahs. The ability of ordinary people to collaborate, to launch new businesses, to invent new medicines, and to provide goods and services of all kinds has never been easier.

Technology is allowing more people to escape the destitution and darkness of poverty so they can live in the incandescent and LED-lit world of modernity. As more people get richer, the competition for land and water, iron ore and petroleum, wheat and soybeans, will continue, just as it always has. This book isn't a blind celebration of technological advancement. Nor is it one that touts a particular method of innovation or even a particular sector. But it does unashamedly celebrate business and entrepreneurs because they are driving the trend toward Smaller Faster Lighter Denser Cheaper.

This book puts a great deal of emphasis on energy and power systems. That focus is purposeful. The energy sector is by far the world's biggest industry, and every sector of the global economy depends directly or indirectly on it. The availability of cheap, abundant, reliable energy is what separates the wealthy from the poor and fuels economic growth. That growth fosters both human liberty and environmental protection. As we go forward, we will need to make energy Cheaper so that more people can join the modern world. We will need more natural gas and more nuclear energy, more oil and solar energy, and yes, more coal.

In Part I, I'll look back at some of the examples of our quest for Smaller Faster Lighter Denser Cheaper and highlight a few of the historical innovations that have changed our lives, including the printing press, the vacuum tube, and digital communications. I will discuss some of the negative outcomes that have come about from, or are unintended consequences of, our innovations. The section concludes with a look at the arguments being put forward by the catastrophists and discusses the pivotal question: should we continue innovating, or retreat to the past?

Part II is a wide-ranging section that examines the push for Smaller Faster Lighter Denser Cheaper in history and in the current day. It looks at the technologies used in the Tour de France as well as those being deployed in education and medicine. It shows how the push for Smaller Faster has motivated industrial giants like Ford and Intel and how those same catalysts are motivating today's start-ups.

In Part III, I dive into the energy sector. Every year, the people of the planet spend roughly $5 trillion on energy.[16] Finding, refining, and delivering the gargantuan quantities of energy needed by the world's consumers requires an epic effort. I show how the energy sector typifies the push for Smaller Faster, and particularly the effort for Cheaper.

In Part IV, I look forward and offer a few ideas as to how we can continue fostering innovation. I explain why, regardless of your beliefs about climate change, the best no-regrets policy for the future is N2N—natural gas to nuclear. I also explain why the United States will dominate our Smaller Faster Lighter Denser Cheaper future.

Now on to Part I, and the project that offers the world's single biggest example of our desire for Faster Cheaper: the Panama Canal.

PART I:

The Push for Innovation,
Its Consequences, and
the Degrowth Agenda

1

PANAMA

DIGGING A FASTER CHEAPER WAY TO TRAVEL

For more than five centuries, humans have been surveying the Panamanian Isthmus in the relentless pursuit of a Faster Cheaper way to travel the oceans. Long ocean voyages are expensive. Wages must be paid. Meals and freshwater must be supplied to passengers and crew every few hours. And the longer a ship stays at sea, the more likely it is to be damaged or sunk by bad weather.

The Isthmus was the logical place to launch an attempt to cut the distance from the Atlantic to the Pacific. If a canal could be completed, a ship going from New York to San Francisco could avoid going all the way around Cape Horn, a months-long voyage of 13,000 miles. A canal could shorten the trip by 8,000 miles. A voyage from New Orleans to San Francisco via an Isthmian canal could save more than 9,000 miles.[1] A canal would mean Faster and Cheaper ocean travel.

The pursuit of Faster Cheaper travel across the Isthmus has been ongoing for the past 130 years. Indeed, the digging continues to this day. During my visit to the Canal Zone in August 2013, I could hear the dynamite blasts being used to deepen and widen the canal. Dredges were actively working in the Culebra Cut, hauling yet more rock out of the narrowest section of the waterway.

In 2014, Panama will celebrate the hundred-year anniversary of the opening of the canal, a celebration scheduled to coincide with the biggest overhaul in the canal's history: a $5.2 billion widening and deepening project that will allow the world's biggest container ships to move between the Atlantic and Pacific Oceans in a matter of hours.

Prior to the expansion, the canal's locks could handle ships that were a maximum of about 295 meters long (968 feet) and 33 meters wide (109 feet). After the expansion, the locks will be able to handle ships that are 366 meters long (1,200 feet) and 49 meters wide (161 feet). For a global shipping industry increasingly reliant on giant container ships, the results will be profound. Before the expansion, the canal could handle vessels carrying up to 5,000 containers; after the expansion, it will be capable of handling ships carrying up to 13,000 containers (known in the business as TEUs).[2] In the ocean-going shipping business, bigger ships usually mean Cheaper.

Building the canal was the moon-shot of the nineteenth century and early twentieth century. No other civil engineering or construction project in modern human history can rival it or even come close in terms of scale, quantity of dirt moved, or number of lives lost in the process. At the time the canal was built, it was both the most ambitious, most expensive and, unfortunately, most deadly, engineering feat ever attempted. There is no exact count of the people who died—the vast majority of them felled by disease—during the entire effort to build the Panama Canal. It may have been as high as 28,000.[3]

The Panama Canal wasn't the first effort at moving lots of dirt to enable more water-borne commerce. In 1761, the Duke of Bridgewater commissioned the Bridgewater Canal, on which coal from the mines in Worsley could be hauled to the city of Manchester. Over the ensuing decades, Manchester became a manufacturing powerhouse. By 1853, it had more than one hundred cotton mills.[4] In 1869, an effort led by French engineers succeeded in connecting the Mediterranean Sea with the Red Sea with the completion of the Suez Canal.*

Emboldened by his success in the desert on the Suez Canal, an uncomplicated sea-level waterway, a pompous French diplomat named Ferdinand de Lesseps convinced himself and numerous French investors that he could repeat his success in Panama, and that he could do so by building yet another sea-level canal. He was wrong. Spectacularly

---

* The Suez route opened to traffic just six months after the opening of another major public works project aimed at providing Faster Cheaper transportation: the Transcontinental Railroad. In May 1869, the last spike was inserted into railroad ties at Promontory Summit, Utah.

June 1909: Afro-Caribbean workers operating air drills in the Culebra
Cut. (Also known as the Gaillard Cut, in honor of the American engineer David D. Gaillard, who managed the excavation of the Cut during
the height of the work on the canal. Gaillard died in 1913, felled by a
brain tumor.) Completing the Cut required the removal of 100 million
cubic yards of dirt and rock.[5] To put that 100 million cubic yards in
perspective: Cowboys Stadium—the palatial $1.3 billion home of the
Dallas Cowboys, which seats 80,000 people—has a volume of 3.85 million cubic yards.[6] Therefore, the material removed from the Cut would
fill Cowboys Stadium 26 times. At the peak of construction, about 6,000
workers were excavating the Cut, filling 160 trainloads of spoil per day.[7]
John Stevens, a dynamic American engineer who headed the canal effort
for several years, wrote that the excavation of the Cut was "a proposi-
tion greater than was ever undertaken in the engineering history of the
world."[8] *Source*: Library of Congress, LC-USZ62–75161.

wrong. The idea of building a sea-level canal in Panama was foolish from the get-go. But it took years of failure and enormous financial losses before de Lesseps and his French backers finally conceded and the Americans took over.

The desire for a Faster Cheaper route through Panama that would allow travelers to easily traverse the continent first arose in the early 1500s, when the Spanish explorer Vasco Nuñez de Balboa succeeded in crossing the Isthmus on foot.[9] By 1811, a German scientist and adventurer named Alexander von Humboldt was declaring that Nicaragua was the best route for a path between the Pacific and the Atlantic. (Nicaragua continues to be discussed as an option for a new canal. In 2013, a Chinese company announced it had been awarded a hundred-year concession that would allow it to build an alternative to the Panama Canal. The project has an estimated cost of $40 billion.)[10]

In 1882, the company that de Lesseps controlled, the Compagnie Universelle du Canal Interocéanique de Panama, began excavating the Culebra Cut. (The word "culebra" is Spanish for "snake.") They optimistically estimated that they would be finished with their excavation by 1885.[11] The French effort to build the canal failed for many reasons. Chief among them was de Lesseps's failure to understand the immensity of the excavation that would be required.

In his landmark book on the building of the canal, *The Path Between the Seas*, historian David McCullough wrote that the variable geology of the Cut was "fascinating terrain to a geologist, but for the engineer it was an unrelieved nightmare."[12] The earth in the region was a mixture of shales, marls, and clays along with some igneous and volcanic rock. The clays were the most problematic because, as McCullough points out, after a heavy rain, they "became thoroughly saturated, slick, and heavy, with a consistency of soap left overnight in water."[13] Numerous landslides forced the engineers to make the Cut wider than they had planned. That was a problem because the nine-mile-long Cut was being made in the saddle between two big hills. As the Cut was widened, more and more dirt, clay, and rock had to be removed. "The deeper the Cut was dug, the worse the slides were, and so the more the slopes had to be carved back," explains McCullough. "The more digging done,

August 2, 2013: A cruise ship heading south through the Culebra Cut. The excavation of the Cut, which began in 1882, was ongoing even as this ship passed. Dredging operations, including the use of explosive charges to break up the rock in the Cut, continued nearly around the clock. The sound of the explosions could easily be heard as far away as Canopy Tower, a popular bird-watching spot located about three kilometers (1.5 miles) east of the Cut. *Source:* Photo by author.

the more digging there was to do. It was a work of Sisyphus on a scale such as engineers had never before faced."[14]

The Cut became known as "Hell's Gorge" due to the dust, heat, and smoke from the coal-fired steam shovels, and nearly constant noise. The working conditions were made worse by the nearly constant danger of dying on the job. Workers were crushed by equipment or falling rock. Others were killed when dynamite accidentally detonated. From start to finish—and there were plenty of interruptions as the French effort faltered—the excavation of the Cut took thirty-one years until the canal was finally opened to traffic.[15]

In many ways, the opening of the Panama Canal on August 15, 1914, marks the true beginning of the twentieth century.[16] It opened just after the beginning of World War I.[17] It opened at about the same time that the internal combustion engine, the automobile, and the airplane were all coming of age—and all of them made transportation Cheaper than ever before. The canal was the first major public works project to utilize electricity on a large scale. The locks were operated by electric motors and switches, all of which were made by an upstart company called General Electric.

Today, a full century after it opened to traffic, the Panama Canal continues to be one of the largest and most astounding feats of human ingenuity on the planet. To transit the canal by boat, or to fly over it in an airplane, is to be awed by the human desire to achieve, to innovate, to go Faster.

The drive toward Smaller Faster Lighter Denser Cheaper that the Panama Canal represents is manifest in many other examples throughout human history, and I'll discuss a few of the most transformative ones in the next chapter. They all have their origins in an innovation engine that has no peer: the human brain.

2

THE TREND TOWARD SMALLER FASTER LIGHTER DENSER CHEAPER

THE BRAIN

The gravimetric power density of the human brain is 100,000 times that of the Sun.*

Yes, it sounds implausible. The Sun is massive. It's the engine for nearly all life on earth. But it is a verifiable fact. My pal Mark Ehsani, an engineering professor at Texas A&M University who heads the school's Advanced Vehicle Systems Research Program, first told me about the power density of the brain in 2010.[1] He walked me through the math. Our brains make up just 2 percent of our body weight, and yet they consume about 20 percent of all the calories we burn.[2] The average power flow in the human body is about 100 watts. Twenty percent of that would be 20 watts. The average brain weighs about 1.5 kilos. Simple division, then, shows that the gravimetric power density in the human brain is approximately 13 watts per kilogram. Meanwhile, the gravimetric power density of the Sun is about 0.00019 watts per kilogram.[3]

The huge difference in power density between the Sun and the brain makes sense when you think about it. The Sun is made up of gases, a big ball of plasma.[4] The brain is a tangled mass of fatty liquid. Water is heavy. Gases are not.

---

* Power density is a measure of the energy flow that can be harnessed in a given area, volume, or mass. I discuss all three types of power density: areal, volumetric, and gravimetric. The book also discusses energy density, which is the amount of energy contained in a given volume or mass.

The brain is not only extraordinarily power dense, it also supports the most complex network in the universe. As Steven Johnson explains in his 2010 book, *Where Good Ideas Come From*, the brain contains about 100 billion neurons. And "the average neuron connects to a thousand other neurons scattered across the brain, which means that the adult human brain contains 100 trillion distinct neuronal connections, making it the largest and most complex network on earth." By comparison, Johnson points out that there are about 40 billion pages on the World Wide Web. "If you assume an average of ten links per page, that means you and I are walking around with a high-density network in our skulls that is orders of magnitude larger than the entirety of the World Wide Web."[5]

The brain has greater power density than the Sun, is more complex than the Internet, and yet is so compact, it can fit inside the confines of a St. Louis Cardinals baseball cap. That's quite a machine. Whether this particular machine was invented by a supreme being or is the result of natural evolutionary processes, it is itself an exemplar of the trend both in nature and society toward density, toward making things Smaller Faster Lighter.

Here are a handful of other historical examples of the trend toward doing more with less.

THE PRINTING PRESS

Sir Francis Bacon (b. 1561, d. 1626) is considered the father of the scientific method, and he named the printing press, gunpowder, and the compass as the most important inventions of his time. In 1620, he wrote that those innovations "have changed the appearance and state of the whole world; first in literature, then in warfare, and lastly in navigation; and innumerable changes have been thence derived, so that no empire, sect, or star appears to have exercised a greater power and influence on human affairs than these mechanical discoveries."[6]

While gunpowder and the compass have undoubtedly changed history, I'm sticking with Bacon on his first choice. The printing press—developed in about 1440 by Johannes Gutenberg—allowed books to

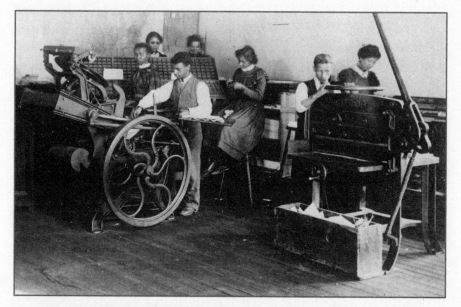

1899: The printing operation at Claflin University, a historically black school located in Orangeburg, South Carolina. *Source*: Library of Congress, LC-USZ62–107845.

be Smaller Lighter Faster Cheaper. Sure, the original Gutenberg Bibles were huge, with each page measuring about 17 inches by 12 inches, but as printers got better at their trade, they developed Smaller fonts and better papers, which allowed books to get Lighter. In the decades following Gutenberg, presses were continually refined so that they printed Faster, and as that printing got Faster, books became radically Cheaper.[7]

The movable-type invention by Gutenberg (b. 1398, d. 1468) changed the world like no other innovation ever has. As historian Abbott Payson Usher explains, the development of printing, "more than any other single achievement, marks the line of division between medieval and modern technology." Printing was among the first instance of "the substitution of mechanical devices for direct hand work in the interests of accuracy and refinement in execution as well as reduced cost."[8] In other words, the printing press enabled Faster Cheaper.

By 1500, more than 2,500 European cities had a printing press.[9] The proliferation of the printing press made education Cheaper. Once reserved only for the rich, the clerics, and the nobility, Cheaper books allowed common people to access knowledge. Gutenberg's invention allowed Faster dissemination of discoveries and scientific information. It increased accuracy. And perhaps most important, it took the control of ideas away from the Catholic Church and gave them to the masses. Without the printing press, there would have been no Renaissance, no Reformation. Martin Luther, the German cleric who lit the fuse on the Reformation, once declared that printing was "God's highest and extremist [sic] act of grace."[10]

Today, thanks to the Internet, billions of people on the planet have access to a virtual printing press; they can instantly publish nearly anything they want to say. If they want to read books, they can download them onto their computer. Project Gutenberg, founded by a visionary named Michael S. Hart, now has more than 42,000 books available for download.[11] Every one of those books is available for free.

THE VACUUM TUBE

The spread of freedom, democracy, and racial integration around the world has many causes. But one that cannot be overlooked is rock and roll, which along with most of the electronic inventions of the twentieth century was a child of the vacuum tube.

Lee De Forest has never occupied the same hallowed place as Thomas Edison in our pantheon of inventors. But by perfecting the vacuum tube, in much the same way that Edison perfected the lightbulb, De Forest helped birth a seminal technology. De Forest's vacuum tube changed music, and in doing so, it changed history. The vacuum tube corralled some of the smallest and fastest things known to humans—electrons— and made them malleable. It put those electrons into the hands of creative people, from Buddy Holly and Chuck Berry to Jimi Hendrix and Bob Marley, who were ready to twist them into entirely new sounds.

In 1906, De Forest, an American, invented the triode vacuum tube. It was the first electrical device that could amplify a weak electrical

Perfected by the American inventor Lee De Forest (1873–1961) in 1906, vacuum tubes (sometimes called electron tubes) can take weak signals and make them stronger, or act as a switch to stop and start the flow of electrons.[12] When heated to somewhere between 1,000 and 2,400 degrees Celsius, a cathode boils off electrons into the vacuum inside the tube. The electrons then pass through a grid, or several grids, which control the flow of electrons before they reach the anode, where they are absorbed. If the cathode, grid, and anode, are properly designed, the tube boosts a small AC current into a larger one, thereby creating amplification.[13] Vacuum tubes were essential to the Information Age. The MANIAC computer built at the Institute for Advanced Study at Princeton, New Jersey—the first computer to use random access memory—used 2,600 of them.[14] *Source:* Photo by author.

signal. Vacuum tubes went into the guts of amplifiers, radio receivers, telephone switchboards, TVs, and nearly every other significant communications device created between 1900 and 1950. The vacuum tube put real power—the wattage needed to be heard at loud volume by large crowds—into the hands of musicians who were ready to, as Jack Black put in *School of Rock,* "stick it to the man."[15]

Duke Ellington, Count Basie, Tommy Dorsey, and other big-band leaders needed a dozen or more players to make a big sound. By contrast, relatively low-cost amplifiers hooked to cheap electronic pickups on mass-produced guitars meant that four musicians, or sometimes even just three, could rock the foundation of nearly any building.

The vacuum tube transformed the guitar from an instrument more suited to the parlor and folk singers into a musical-cultural icon that has come to represent youth and rebellion. Armed with a Fender Telecaster—the world's first commercially successful solid-body electric guitar, introduced in 1950, or another iconic instrument like the Gibson Les Paul (1952) or the Fender Stratocaster (1954)—and an amplifier made by Fender, Vox, or Marshall, a single musician could hold his own against the biggest of the big bands.[16] The guitar democratized the making of music. The guitar didn't require the years of intense training required by more demanding instruments like the violin, clarinet or saxophone. Bob Dylan and a host of other singer-songwriter-rock-and-rollers made their livings with just three or four basic chords. The electric guitar allowed a talented musician like a Hendrix, Eric Clapton, or Freddie King or even untalented ones (a list too long for this book) to bend the minds of tens of thousands of listeners from Wembley Stadium to the Cotton Bowl.

Thanks in large part to the ingenuity of a California radio repairman named Leo Fender, rock and roll gained the tools it needed. Fender used his knowledge of vacuum tubes and electronics to start building guitars, amplifiers, and basses at his shop in the Los Angeles area. Fender's designs were quickly adopted by musicians like Muddy Waters, Lionel Hampton, Buddy Guy, Keith Richards, Bruce Springsteen, Stevie Ray Vaughan, and ultimately, millions of others.[17] Cheap vacuum tubes, which were followed by even Cheaper integrated circuits (which could perform the same functions) allowed musicians more flexibility and tonal range than had ever been imagined.

The vacuum tube birthed rock and roll and set the stage for the Information Revolution. In doing so, it changed the world by making music a global commodity—one that connected people of different cultures, economies and languages by giving them a common lyric and a common beat.

On February 9, 1964, the Beatles made television history by appearing on the *Ed Sullivan Show*. Paul McCartney played an electric bass. George Harrison and John Lennon played electric guitars. That appearance, watched by an astounding 40 percent of the US population, launched what became known as the British Invasion.[18] That transfer of musical styles—all of it made possible by the vacuum tube—helped rock and roll become a global phenomenon. The Beatles' appearance on the Sullivan show "opened the transatlantic floodgates," writes Tim Brookes in *The Guitar: An American Life*. After the Beatles, came other British groups: the Kinks, the Moody Blues, the Who, and, of course, the Rolling Stones. That motley group of Limeys—the Stones in particular—introduced white American audiences to the black American music that had inspired them. "Perhaps the most important contribution of the British Invasion was in helping America connect with its own past and its alienated present," writes Brookes.

The advent of rock and roll—which included the success of the Beatles, along with that of black blues artists, and southern singers like Elvis Presley, who was born in Tupelo, Mississippi—undermined long-held prejudices and helped the United States become more integrated. As Brookes points out, rock and roll held a giant mirror in front of Americans and allowed them to see Jimi Hendrix and Freddie King not as black men but as dynamic musicians.

When the Beatles came to America, the Fab Four were asked by an interviewer about what they wanted to see during their visit. They quickly answered "Chuck Berry and Bo Diddley," the great African American electric-guitarists and performers. When the interviewer didn't recognize the two names, John Lennon's "indignation flattened the guy." Lennon asked, "Don't you even know your own music?"[19]

Rock and roll did as much, or more, to bring down the Berlin Wall as any other single factor. In 2003, Mikhail Safonov, a researcher at the Institute of Russian History in St. Petersburg, wrote a piece for *The Guardian*, in which he declared that it was John Lennon who "murdered the Soviet Union." Safonov wrote that the history of the Beatles' persecution in the Soviet Union—their music was banned and the group was prohibited from playing there—was "the history of the self-exposure of the idiocy of Brezhnev's rule. The more they persecuted something the world had already fallen in love with, the more they exposed the falsehood and hypocrisy of Soviet ideology."[20]

How the Beatles undermined the Iron Curtain was the subject of the 2009 documentary *How the Beatles Rocked the Kremlin*. The Beatles woke up "an entire generation of Soviet youth, opening their eyes to 70 years of bland official culture and rigid authoritarianism."[21] Created by the veteran British filmmaker Leslie Woodhead, the documentary contains numerous interviews with now-middle-aged Russians who discuss the importance of the Fab Four. One of them says simply, "It's all thanks to the Beatles. They helped destroy the Evil Empire."[22] (In 2013, Woodhead released a book with the same title as the documentary.)[23]

The Soviet authorities weren't alone in worrying about rock and roll. In 1964, the Israeli government refused a request to have the Beatles play in that country after the group was deemed "liable to have a negative influence on the youth."[24] In 1975, East German authorities prohibited the musicians who belonged to the Klaus Renft Combo, a rock and roll group, from performing, telling them that the lyrics to their songs "had absolutely nothing to do with socialist reality . . . the working class is insulted and the state and defense organizations" had been "defamed." Rather than stick around, one member defected to the West. Other members of the group were briefly imprisoned by the East German authorities.[25] In the 1980s, East German authorities also banned the British punk rock group The Clash.[26]

The outlawing of rock and roll groups didn't end with the fall of the Iron Curtain. In 2012, members of Pussy Riot, an all-female punk rock group, were jailed in Russia after they performed a demonstration against the country's strongman-president Vladimir Putin at Moscow's main Orthodox cathedral. Three members of Pussy Riot were convicted and imprisoned on charges of "hooliganism motivated by religious hatred" even though their antics in the church were plainly aimed at Putin's repressive government.[27] It's remarkable that Putin and his band of Kremlin-based kleptocrats are so threatened by a group of young women armed with nothing more than Fender Stratocasters.

The vacuum tube allowed musicians to be heard as individuals, and in doing so liberated millions of people. Lee De Forest, the

Alabama-born inventor who perfected the vacuum tube, would eventually win more than three hundred patents.[28] But none of his other inventions would ever be as important as the vacuum tube.

THE AK-47

Mikhail Kalashnikov made killing Cheaper. That's hardly an achievement for which most people would want to be known. Nevertheless, Kalashnikov, a former tank mechanic for the Russian military who died in December 2013, deserves a place in history for designing the AK-47, a weapon that one writer has called the "most effective killing machine in human history."[29] Kalashnikov's design was effective because it was Smaller Lighter Cheaper than other assault rifles.

In his 2010 book, *The Gun*, C. J. Chivers, the sharp-eyed war correspondent for the *New York Times*, described the key attributes of the AK-47, "shorter and lighter than traditional rifles but larger than submachine guns." The AK-47 "could be fired either automatically or a single shot at a time. Their smaller, intermediate-power cartridges allowed soldiers and guerrillas to carry more ammunition into battle than before." In addition to the increase in firepower, the rifle was "an eminently well designed tool—reliable, durable, resistant to corrosion, and with moderate recoil and a design so simple that their basics could be mastered in a matter of hours."[30]

This entry could be devoted to firearms in general, as the development of firearms changed the balance of power among nations. The mass production of firearms, which began in earnest in the early nineteenth century, was a driving force during the early days of the Industrial Revolution. The need to produce large quantities of precisely machined parts led to major advances in manufacturing techniques that quickly spread to other industries. Therefore, any number of other firearms, including the Kentucky rifle, the Gatling gun, or the Colt M1911 .45 caliber semi-automatic pistol, could be listed here as a game-changing weapon.[31]

But I'm sticking with the AK-47 because of its ubiquity and price. Since it was developed in about 1947, as many as 100 million Kalashnikov rifles (both the AK-47 and AK-74) have been produced. (The

The AK-47. *Source:* Wikipedia.

American-made M-16 is a relative laggard, with about eight million copies). In 2006, Amnesty International reported that in some parts of Africa, an AK-47 could be purchased for as little as $30.[32] In addition to its low cost, the AK has gained renown for its simplicity and ability to fire under almost any conditions. The rifle has only nine parts and can fire up to six hundred rounds per minute.[33] Numerous videos available on YouTube show that the Kalashnikov can be fouled with water, dirt, leaves, and other debris, and yet it still operates.[34]

In 2005, the BBC called the Kalashnikov "an icon of violence in the 20th Century."[35] The outline of the AK with its distinctive curved magazine is on the flag of Mozambique as well as the flag of Hezbollah, the Shiite militant group that has long been backed by Iran.[36] In *The Gun*, Chivers deems the AK-47 as a "stubbornly mediocre" firearm.[37] That may be true. But that mediocrity has almost certainly resulted in hundreds of thousands, or perhaps even millions, of deaths.

Politicians and terrorism experts often focus on the risks associated with weapons of mass destruction, including ones that are chemical, biological, or atomic. But firearms like the AK-47 are the real killers. Up to 90 percent of all civilian casualties in conflict zones are caused by small arms like the AK-47. (The definition of small arms includes assault rifles, pistols, mortars, landmines, and grenades.)[38] By some estimates, small arms are involved in more than a thousand deaths every day.[39] Of course, there's no way to know how many of those deaths can be attributed to the use of the AK-47. But as one of the most common of all small arms, Kalashnikov's rifle has surely resulted in enormous human losses.

THE HABER-BOSCH PROCESS

There will always be arguments as to which invention is the most significant. But when it comes to basic human survival and the ability of people to have sufficient food on their tables, the Haber-Bosch process stands alone. As one author put it, no invention has "had such an impact on our civilization as did the synthesis of ammonia from its elements."[40]

In the Haber-Bosch process, natural gas and atmospheric nitrogen are converted into nitrogen fertilizer. To understand the importance of the process requires a modicum of history, chemistry, and math. During the late 1800s and early 1900s, farmers were desperate for more nitrogen-based fertilizers because nitrogen is an essential plant nutrient. The problem was that the world's primary source of raw material for fertilizer production was a large deposit of guano (bird poop) located on the Chilean coast. (Guano was also retrieved from other sources, including local bird roosts and bat caves.) Mining and hauling the guano from such a remote location presented many logistical problems, which made fertilizer expensive.

Two Germans, Fritz Haber and Carl Bosch, made fertilizer Cheaper by inventing a method of manufacturing that pulls nitrogen out of the atmosphere and combines it with hydrogen atoms that are usually derived from natural gas (CH_4). The process, for which Haber won a patent in 1911, uses high temperature, about 500 degrees C, as well as high pressure (about 200 times normal atmospheric pressure), and an iron catalyst. The product is ammonia (NH_3), a substance that is superior to guano when used as a raw material for fertilizers. It's also essential to the production of nitric acid, which is used in the production of explosives. That last fact undoubtedly explains why Haber and Bosch had to wait years for proper recognition. Haber was awarded the Nobel Prize in chemistry in 1918 "for the synthesis of ammonia from its elements."[41] Bosch won the same Nobel award in 1931.[42] While the importance of

World Fertilizer Use and Grain Production, 1961–2011

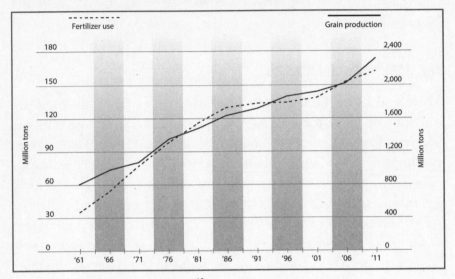

Source: Earth Policy Institute.[43]

their invention was not questioned, the two were also blamed for giving Germany the capacity to produce more explosives, and therefore prolonging World War I. While the history and chemistry are important, it's simple math that explains why the Haber-Bosch process is so important. About two out of every five people on earth are now getting the protein in their diets thanks to the Haber-Bosch process.[44]

The dramatic increases in global grain production that have occurred over the past few decades are a direct result of the Haber-Bosch process. As the graphic above shows, these increases have occurred in tandem with increasing use of fertilizer.

The father of the Green Revolution, Norman Borlaug, fully understood that higher productivity on farmland was due to fertilizers produced with the Haber-Bosch process. In 1970, in the speech he gave while accepting the Nobel Peace Prize, Borlaug declared, "If the high-yielding dwarf wheat and rice varieties are the catalysts that have ignited the Green Revolution, then chemical fertilizer is the fuel that has powered its forward thrust."[45]

THE DIESEL AND THE JET TURBINE

Regardless of where you travel on this planet, it's unlikely you'll be very long out of earshot of the familiar rattle of a diesel engine. Flying anywhere on a commercial airline almost certainly comes with the familiar whine of a jet turbine. Together, the diesel and the turbine have made transportation Faster Cheaper.

In his 2010 book *Prime Movers of Globalization: The History and Impact of Diesel Engines and Gas Turbines*, Vaclav Smil declares that those two machines are "more important to the global economy than are any particular corporate modalities or international trade agreements."[46] Smil continued, writing:

> The human quest for a higher standard of living, profits, and power and the human propensities for long-distance trade and exploration have been the key motivating forces. But without the two prime movers [the diesel and turbine], trade would not have achieved its truly planetwide scope or have done so at such massive scales, at such rapid speed, and at such affordable costs.[47]

The centrality of diesel engines to the modern economy can be demonstrated by one fact: more than 80 percent of all the freight moved in the United States is conveyed on machines powered by diesel engines.[48] The key advantage provided by the diesel engine is its efficiency, which is 25 to 40 percent higher than comparable gasoline engines that use spark-ignition systems.[49] Some of that efficiency comes from the higher energy density of diesel fuel, which contains about 17 percent more heat energy by volume than gasoline.[50] But it's also true that the engine's creator, Rudolf Diesel, born in Paris in 1858, was consumed by the desire to create engines that were Cheaper to operate.[51] While in school at the Munich Polytechnic, Diesel learned that only about 10 percent of the heat energy used by steam engines was turned into useful work. He saw an opportunity. Diesel wrote that his desire to create a more efficient engine "dominated my existence. I left the school, went into practice, had to win a position in life. The thought pursued me incessantly."[52]

By 1897, Diesel wrote that he had created "a thoroughly marketable machine."[53] As we now know, Diesel succeeded. Or rather, his belief in the need for an efficient, compression-ignited, internal-combustion engine did. On a personal level, Diesel was ruined. By 1913, he was heavily in debt and distressed by criticism from colleagues who claimed his engine wouldn't work. In September 1913, while aboard a ship crossing the English Channel, Diesel apparently jumped overboard. His body was found about two weeks later.

Although Diesel didn't live to reap the rewards or the accolades, his name has become synonymous with motive power. Some of the world's biggest engines use Diesel's idea. Finland-based Wärtsilä is now selling diesel engines that weigh about 2,100 metric tons and have power outputs of more than 94,000 horsepower (70 megawatts), which are for use in large container ships.[54] While those numbers are certainly Bunyanesque, those ultra-large engines are also among the most efficient ever built, with thermal efficiencies of 50 percent or more.

While diesels are driving surface-based trade, jet turbines have made global air travel into a routine experience. Six decades ago, passenger airliners relied on piston-driven engines that used high-octane gasoline. In the 1950s and 1960s, piston-driven airliners gave way to jetliners. Jet aircraft became dominant because they can fly about three times as fast and twice as high as their gasoline-powered cousins. That means that passengers can save huge amounts of time and do so while flying in the upper reaches of the troposphere, which is usually above the levels where weather and air turbulence is a problem.[55]

The effect of the jet turbine can be seen by looking at the astounding growth in air travel. In 1950, the total volume of global air travel measured in passenger-kilometers was 28 billion.[56] By 2011, that figure had increased to 5.2 trillion passenger-kilometers, a 186-fold increase.[57] Today, we take for granted the ability to fly to Paris, New York, and Guayaquil. And while we curse the crowded airplanes and the sometimes-grumpy flight attendants, it's easy to forget just how much Cheaper and more convenient air travel has become. In 1946, TWA offered flights between New York City and Paris for $675 per person.[58] In today's money, that fare would be close to $8,000.[59] That's a huge sum of money considering that a recent search on Orbitz found half a dozen airlines offering fares of less than $1,000 for a round-trip ticket from Newark International to Charles de Galle airport in Paris.

Today's flights are not only Cheaper, they're also Faster. Back in 1946, one of the most popular long-range aircraft was the Lockheed Constellation, which was powered by four large piston-driven engines. The trip from New York to Paris took about twenty hours, with two stops for refueling. Today, that same trip is done nonstop and takes about eight hours. Nevertheless, the twenty-hour travel time must have been attractive in those pre–Jet Age days, as it was about five days Faster than traveling the same route by ocean liner.[60]

The history of the diesel and turbine reflects the never-ending quest for Smaller Faster Lighter Denser Cheaper.

Conventional piston-driven engines generally rely on combustion that happens in four precisely choreographed stages. These four-stroke internal combustion engines—which are also known as Otto cycle (for the German inventor Nikolaus Otto)—are found in everything from automobiles to lawnmowers. The devices operate in four stages: intake, compression, power, and exhaust. Jet turbines are different in that they allow concurrent and continuous combustion. The introduction of fuel, along with compression, ignition, combustion, and exhaust of hot gases, occurs continuously in different sections of the machine. This design, also known as the Brayton cycle (named after American inventor George Brayton, who was born in 1830 in Rhode Island), was first tested in military aircraft in 1939. The first combat airplanes to use the turbines went into service in 1944, and the first turbine-powered commercial aircraft, the British Comet, began carrying passengers in 1952.[61] At that time, the thermal efficiency of the turbines (that is the amount of heat energy turned into useful work) was about 18 percent.[62] Today's turbines are far more efficient. And while efficiency is certainly important, turbines are also Denser than piston engines; that is, they have higher gravimetric power densities. The ongoing push for higher power-to-weight ratios can be seen by looking at aircraft engines.

In 1903, when Orville Wright changed history with his short flight aboard the Wright Flyer at Kitty Hawk, North Carolina, he and his

Cheaper Airfares: The Declining Cost of US Domestic Airfares, 1979–2011

(In constant dollars, fees included)

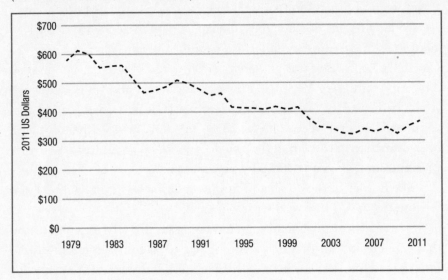

Source: Airlines for America.[63]

brother Wilbur were relying on an engine that produced 116 watts per kilogram. Forty-two years later, in 1945, the Boeing B-29 Superfortress used to drop the atomic bombs on Hiroshima and Nagasaki was powered by four giant air-cooled radial engines (the Wright R-3350), each of which had gravimetric power densities of about 1,354 watts per kilogram.[64] In other words, the power densities of B-29's engines were nearly 12 times greater than that of the Wright Flyer engine. By the 1950s, commercial jetliners were using turbines with gravimetric power densities of about 4,000 watts per kilogram, or 34 times greater than what had propelled the Wright Flyer.

Today, thanks to advances in computer modeling, fluid dynamics, carbon-fiber composites, and other manufacturing techniques, the pinnacle of jet turbine design may be General Electric's GEnx-1B, which powers the Boeing 787. The new turbine's gravimetric power density is nearly 15,000 watts per kilogram. That's about 130 times the power density of the Wright Flyer engine. In addition to its enormous power

The GEnx-1B is among the latest designs in aircraft turbines. Its power density is about 15,000 watts per kilogram. *Source*: General Electric.[65]

density, the GEnx-1B is about 30 percent quieter than the turbine it is replacing, and it consumes about 15 percent less fuel.[66] (For more on gravimetric power density from humans and horses to steam engines and jet turbines, see Appendix C.)

Today, jet travel has become so routine as to be almost boring. On an average day in 2011, some 7.6 million people boarded commercial airliners. By 2016, the airline industry expects that number to climb to 9.8 million people per day.[67] Of course, the jetliners crisscrossing the globe are carrying more than humans—there's fresh fruit, flowers, mail, and dozens of other types of cargo. At those planes' destinations, the vast majority of that cargo will be unloaded onto diesel-powered trucks.

The diesel and the jet turbine made travel Faster Cheaper and in doing so brought us modernity, along with pollution and sprawl. Diesel trucks and stationary diesel engines belch millions of tons of air pollutants and carbon dioxide into the atmosphere every year. Jetliners are allowing us to travel Faster, but in doing so, they are also allowing the spread of disease. An outbreak of cholera that began in Haiti in late 2010 killed more than 7,500 people. The disease, which hadn't been seen in Haiti in more than a century, was traced to a camp of UN-assigned Nepalese soldiers who had flown to Haiti.[68] The Nepalese were housed in a camp that had inadequate latrines. Feces from the latrines leaked into the Meye River, which in turn, flows into Haiti's main waterways. From there, the disease spread throughout the country, which is plagued by inadequate sewerage and freshwater-distribution systems.[69]

The diesel and the jet turbine, like many other innovations, have brought us great convenience, and in doing so have exacted a price. Nevertheless, the diesel and jet turbine have helped conquer the tyranny of distance; they made the global economy just that—global.

THE TELESCOPE AND MICROSCOPE

The telescope brought the distant nearer. The microscope made the tiny larger. They were the first true extensions of the most important of the human senses. Together, they were the pivotal instruments of the Scientific Revolution. The telescope destroyed the myth of a geocentric universe; the microscope hinted at the nanogalaxies inside cells and molecules. Armed with them, astronomers and physicians could peer into worlds that had never been imagined.

We use those same devices today—albeit far more powerful ones—to peer light-years into space and down to the angstrom level of the atom.* In short, the telescope and the microscope made magnification Cheaper. They allowed ordinary people to see celestial and microscopic phenomena for themselves.

---

* An angstrom is one-tenth of a nanometer, or 0.1 billionth (10^{-10}) of a meter. The unit is named for the Swedish physicist Anders Jonas Ångström (1814–1874), who did pioneering work in spectroscopy. See: http://en.wikipedia.org/wiki/Angstrom.

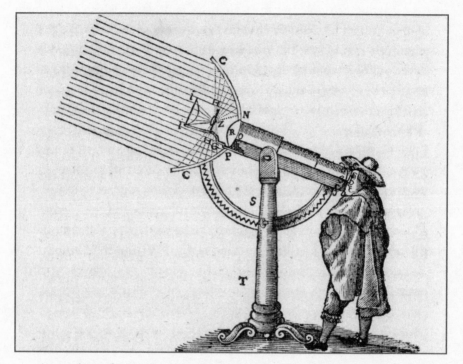

A woodcut of a man looking through a telescope. This image was published in 1637 in "La Dioptrique," an essay by Descartes. *Source*: Library of Congress, LC–USZ62–110450.

For centuries, humans have been making ever more powerful devices in order to see things that are Smaller and more distant. That pursuit blossomed with Galileo Galilei (b. 1564, d. 1642) and Antoni van Leeuwenhoek (b. 1632, d. 1723), the father of microscopy. Obvious examples of continuity are the 11-meter-wide mirror of the Hobby-Eberly Telescope in Fort Davis, Texas, and the University of Manchester's "microsphere nanoscope," which can examine objects as small as 50 nanometers across.[70] (A nanometer is one-billionth of a meter.)

Galileo wasn't the first to use a telescope. But the improvements that he made to that device allowed him to use the telescope as a weapon in the war between reason and faith. In 1610, after months of observations using a 20-power telescope for which he had ground his own lenses, Galileo published his *Sidereus Nuncius (The Message from the Stars),* a forty-page pamphlet.[71] It was the first scientific treatise based on telescopic observations.[72] Arthur Koestler in his history of astronomy, *The Sleepwalkers,* declared that Galileo's "short but momentous book . . . heralded the assault on the universe with a new weapon, an optic battering ram, the telescope."[73]

The knowledge that came from Galileo's optic battering ram made him a target for the Catholic Church and the Inquisition. In 1633, the Church forced Galileo to recant his belief in the Copernican theory of the solar system—with the Sun, not the Earth at its center—a move that saved Galileo from being burned at the stake. (In 1992, Vatican officials finally admitted that Galileo had been right.)[74] While the Church may have had the muscle to force Galileo to recant, it couldn't stop the Scientific Revolution, nor prevent others from confirming his work.

Since the days of Galileo, a "telescope race" has continued unabated as astronomers have sought more powerful devices. Galileo himself relied on telescopes that were about 1 to 1.2 meters (3 to 4 feet) in length.[75] By the 1670s, the Polish astronomer Johannes Hevelius had built a telescope 140 feet long.[76] Since then, designers have come up with a variety of ways to increase the power of telescopes while decreasing their cost. A similar push for Cheaper magnification occurred in microscopy.

Van Leeuwenhoek wasn't the first to build a microscope. But like Galileo, he made critical improvements to the device. Also like Galileo, he relied on himself to grind and polish his own lenses. While other microscopes of the seventeenth century could provide magnification of about 50x, van Leeuwenhoek was able to achieve magnifications of about 270x.[77] That magnification allowed him to be the first to see and describe bacteria, muscle fibers, capillaries, and single-celled organisms. The microscope allowed scientists to study the structure of cells and examine microorganisms. Today, the microscope is commonly used in medical clinics and offices all over the world to examine specimens and help diagnose illnesses.

The American photographer Gordon Parks took this photo of students using microscopes at Bethune-Cookman College in 1943.
Source: Library of Congress, LC-USW3–017132-C.

Nearly three hundred years after van Leeuwenhoek's death, consumers can easily purchase microscopes that are twice as powerful as those he used, but spend only $100 or so. A telescope with twice the magnification of the ones used by Galileo can be had for half that sum.[78] By making magnification Cheaper, the telescope and the microscope democratized science and learning. By bringing the distant near and making the small large, the telescope and microscope liberated the human mind from the intellectual straitjackets of the Church. Thus, they provided the foundation for much of modern science and modern society.

THE PEARL STREET POWER PLANT

In just one year, 1882, Thomas Edison completed 106 successful patent applications. Ponder that for a moment. The great inventor died in 1931 at age 84.[79] In his lifetime, he was awarded 1,093 US patents.[80] Thus, in a single year, 1882, the Wizard of Menlo Park obtained nearly 10 percent of all the patents he would accumulate over his lifetime. No other inventor in US history—with the possible exception of Ravi Arimilli, a researcher for IBM who claimed 78 patents in 2002—has come close to the single-year record that Edison set.[81] His patent record is akin to Joe DiMaggio's 56-game hitting streak in baseball in 1941 or the 100 points that Wilt Chamberlain scored in a single basketball game back in 1962.[82]

On September 4, 1882, Edison began providing electricity from a coal-fired facility located at 255–257 Pearl Street near the southern tip of Manhattan. In doing so, he almost single-handedly created modernity. The 600-kilowatt generator was the world's first centralized power plant, and it sparked a wave of electrification that continues to this day.

Edison's Pearl Street plant made lighting Cheaper. Author David E. Nye explains that Edison's electric lights were "unlike all previous lights, whether candles, oil lamps, torches, fires, or gas mantles. Light by definition had always implied consumption of oxygen, smoke, flickering, heat, and danger of fire. For all of human experience light and fire had been synonymous." With his incandescent bulbs, Edison provided light that was "at once mild and intense, smokeless, fireless, steady, seemingly inexhaustible . . . The enclosed light bulb seemed an impossible paradox. Fire and light would never again be identical."[83]

Electricity is the fire of the nineteenth, twentieth, and twenty-first centuries. Electricity has changed human society like no other form of energy. Edison's breakthrough designs at the Pearl Street plant allowed humans to reproduce the lightning of the sky and use it for melting, heating, lighting, precision machining, and a great many other uses. Electric lights meant workers could see better and therefore make more precise drawings and fittings. Electricity allowed steel producers to operate their furnaces with greater precision, which led to advances in metallurgy. Electric power allowed factories to operate drills and other

Thomas Edison in an undated photograph. *Source*: Library of Congress, Reproduction Number: LC-USZ62–78996.

precision equipment at speeds unimaginable on the old pulley-driven systems, which relied on waterwheels or steam power. As Henry Ford wrote in 1930, without electricity "there could be nothing of what we call modern industry." Electricity, he said, "emancipated industry from the leather belt and the line shaft."[84]

Edison understood the importance of the Pearl Street endeavor, later calling it "the biggest and most responsible thing I had ever undertaken . . . Success meant world-wide adoption of our central-station plan."[85] By 1890, just eight years after Edison launched the beginning of the new world, there were a thousand central power stations in the United States, and new ones were being added at a frenzied pace.[86] Edison's coal-fired power plant was directly responsible for the construction of thousands of others, and that building boom continues to this day.

In late 2013, Maria Van der Hoeven, executive director of the International Energy Agency, remarked on the soaring use of coal, saying that coal is "really emerging as a fuel of choice because of its abundance and affordability." Between 2010 and 2015, the countries of the world are expected to add 285 gigawatts of new coal-fired electric generation capacity. For comparison over that same time period, just 20 giga-watts of nuclear capacity is expected to be built.[87] The proliferation of coal-fired electricity that has given us access to lighting and countless other technologies has also caused a dramatic increase in coal mining, which has taken a heavy toll on miners, particularly those who mine underground. Although coal-combustion technologies have improved since Edison's day, burning coal to produce electricity also releases huge quantities of air pollutants and heavy metals. Those pollutants have taken an additional toll. Coal-fired power plants are among the world's biggest producers of carbon dioxide, the gas that contributes to climate change.

It's readily apparent that the electricity revolution that started on Pearl Street in 1882 has come with a cost. But it's also easy to forget the benefits. Indeed, we've largely forgotten just how awe-inspiring artificial lighting can be. In April 1880, two years before Edison began operating the Pearl Street plant, the town of Wabash, Indiana, arranged to have four big arc lights—each with 3,000-candle power—set up at the courthouse. At that time, artificial light was rare, and promoters traveled the country to demonstrate the power of their arc lights. The excitement preceding the event was so great that special trains were arranged to help carry some 10,000 visitors, along with reporters from forty newspapers, into the Indiana town.

As darkness fell on the settlement on that spring night in 1880, a reporter for the *Wabash Plain Dealer* explained that the town "presented a gloomy uninviting appearance."[88] When the arc lights were switched on, the flood of light should have "caused a shout of rejoicing from the thousands who had been crowding and jostling each other in the deep darkness of the evening." Instead, "No shout, however, or token of joy disturbed the deep silence which suddenly enveloped the onlook-ers." The crowd "stood overwhelmed with awe, as if in the presence of the supernatural. The strange weird light exceeded in power only by the sun, rendered the square as light as midday . . . Men fell on their knees, groans were uttered at the sight, and many were dumb with amazement."[89]

We are no longer "dumb with amazement" at electric lights as the people of Wabash were back in 1880. But we must recall that the flow of cheap, abundant, reliable electrons began on Pearl Street. Modernity began in 1882, when Edison was collecting a new patent every three and a half days.

THE ROLLER-CONE DRILL BIT

Without Howard Hughes Sr.'s roller-cone drill bit, Henry Ford's Model T—along with the entire Age of Automobiles—would have run out of gas.

It's become accepted wisdom that when Ford began mass producing the Model T in 1908, he revolutionized the automotive industry and our transportation system. What's often overlooked is the critical role played by the roller-cone drill bit, an innovation for which Hughes and partner, Walter Sharp, filed for a patent on November 20, 1908, just weeks after Ford began production of the Model T.[90] (Production started on October 1 of that year.)[91]

More than a century ago, long before bits and bytes—described in all manner of tera, giga, mega, and kilo—we had the fishtail drill bit. And it wasn't worth a darn at boring holes in the earth. The business end of the device did look somewhat like a fish's tail or the business end of a very wide screwdriver—a solid piece of steel with curved, sharpened edges. But the fishtail bit's limitations were many. The bits tended to wander off course and couldn't drill effectively in hard-rock formations. Whenever it struck hard rock, the bit would dull quickly, and crews would have to pull the entire drill string out of the well and replace the bit—a costly and time-consuming process. Those limits meant that wildcatters were limited to looking for oil deposits that lay close to the surface. For instance, in 1901, one of the most famous oil wells in history—the gusher at Spindletop, located just outside Beaumont, Texas—came in. That well was drilled to just 1,160 feet.[92] At that time, prospectors looking for oil in Texas and elsewhere were only interested in locations where oil could be found relatively close to the surface. Drilling deeper than 1,000 feet or so simply took too long and cost too much.

The breakthrough came seven years after Spindletop, when Sharp and Hughes—the father of Howard Hughes Jr., the eccentric, reclusive playboy who loved fast airplanes and Faster women—introduced their new roller-cone design, which was vastly superior to the fishtail. Instead of scraping rock as the fishtail bit did, the roller-cone mechanism chipped, crushed, and powdered the rock. That allowed the cuttings from the well to be easily removed by drilling fluid. The bit was also easier to control in the well and had less of a tendency to deviate. Early tests proved the roller-cone bit's superiority. On a well drilled in Humble, Texas, a crew using a fishtail bit was able to bore just 38 feet over nineteen days, or 2 feet per day. When the same crew used one of Hughes's new roller bits, they were able to drill 72 feet in six days, or 12 feet per day.[93]

By making drilling Faster, the roller-cone bit revolutionized the oil and gas sector. Without it, there simply would not have been enough oil production, and therefore enough gasoline, to fuel all of the cars that Ford and other automakers were building. The history of US oil production throughout the 1890s and the first decade of the 1900s shows that production growth was painfully slow. In the decade from 1890 to 1899, production grew only slightly, from 126,000 barrels to just 156,000 barrels per day. By 1909, when Hughes was granted a patent for his design, US oil production had grown to 502,000 barrels per day. A decade later, it had doubled. By 1929, it had doubled again. Forty years later, in 1969, when Neil Armstrong walked on the Moon, domestic production of oil was 9.2 million barrels per day—18 times as large as it was in 1909.[94]

In 2009, the American Society of Mechanical Engineers named the Hughes two-cone drill bit a "historic mechanical engineering landmark." The group said that Hughes's bit "and the rotary drilling system were pioneering inventions that paved the way for the development of technologies and processes still used in the oil field today."[95]

The drill bit didn't just fuel the growth of auto manufacturers like Ford. Nor did it only enrich oil producers and refiners. By making

Fishtail drilling bits like this one dominated the drilling sector in the nineteenth and early twentieth centuries. They were quickly cast aside in favor of roller-cone bits. *Source*: Wikimedia Commons.[96]

drilling Faster, the roller-cone drill bit liberated city dwellers. As author Edward Tenner has written, the automobile came "to represent independence from the rich." With cheap cars and cheap gasoline, Tenner points out that people were liberated from the railroads, streetcar companies, and "center-city landlords. By the 1950s and 1960s, the automotive industry had come to represent big business at its most arrogant, but motorization won because it rallied so many small businesses. Diffuse interests were its political strength." The Automobile Age helped create thousands of small and large businesses. More automobile sales required more auto dealers, mechanics, gasoline retailers, and tire shops. More mobility meant that consumers didn't need to rely on landlords in the city; instead they could move to suburban homes and have a yard of

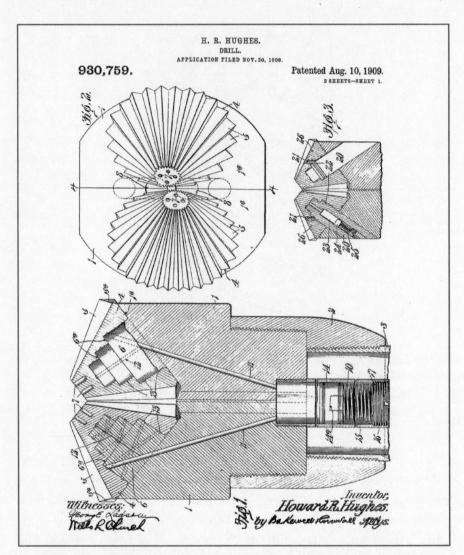

Patent document for the roller-cone bit. Howard Hughes Sr., along with his partner, Walter Sharp, changed the oil and gas industry with their design for the roller-cone drill bit. The success of the drill bit at developing more oil went hand in hand with the growth of the American auto industry. In 1912, after Sharp's death, Hughes bought out his interest. His company, Hughes Tool, is now part of Baker Hughes Incorporated, one of the world's largest oilfield services firms. *Source:* US Patent Office.

their own. That suburbanization led to the building of new roads, more houses, and retail establishments of all kinds. And those businesses helped foster yet more economic activity.

The roller-cone drill bit made gasoline and other refined oil products Cheaper and more abundant. In doing so, it also fostered suburban sprawl, air pollution, and the dreaded time-wasting commute. The roller-cone bit led indirectly to accidents like Deepwater Horizon in 2010, which resulted in a massive oil spill in the Gulf of Mexico. Oil, and its production, continue to be points of conflict in numerous countries around the world, with Nigeria, Libya, and Iraq being obvious examples. Juan Pablo Perez Alfonso, a Venezuelan who was one of the founders of OPEC, famously called oil "devil's excrement." He also claimed "oil will bring us ruin."

No other commodity inspires more passion, or more witless hyperbole, than petroleum. Nevertheless, the world runs on oil. It will continue running on oil for decades to come because petroleum is such a useful commodity. And the roller-cone bit played a key role in making that commodity Cheaper. By making oil Cheaper, the Hughes drill bit allowed huge improvements in living standards around the world, and that trend continues to this day.

DIGITAL COMMUNICATIONS

Back in 1620, Francis Bacon enthused about the printing press and its effect on humanity. Given his enthusiasm for that technology, it's fun to consider what the great scientist might have said about the World Wide Web.

The printing press enabled the leader of the Reformation, Martin Luther, to "throw a lot of ink at the devil."[97] Today, thanks to the virtual printing press that's available via the Internet, nearly everyone can throw ink at his or her devil. And they can do so for free, or nearly so. The importance of today's digital communications networks can scarcely be overstated, particularly for people who are struggling against repressive governments. As one Syrian political activist told the *New York Times* in 2011, "If there's no Internet, there's no life."[98]

The Internet, mobile phones, text messaging, e-mail, GPS, and other digital communications technologies have dramatically changed our society. Thousands of books have been written about the Internet. This book could focus solely on the innovations of digital communications. But instead I will just make an obvious point: while our increased connectivity has brought us tremendous convenience, it has come with a worrisome downside. Our love of ubiquitous computing—the ability to get nearly any type of information we want on our mobile phones at any time—has enabled what Bruce Schneier calls "ubiquitous surveillance."

In April 2013, Schneier, a security technologist and author of a dozen books on privacy and security issues, wrote that surveillance has become "efficient beyond the wildest dreams of George Orwell."[99] (Orwell was the author of the dystopian novel *1984*, published in 1949, which looked at a tyrannical society of the future that was ruled by Big Brother.)

Two months after Schneier's essay was published, Glenn Greenwald, a journalist who was then working for London's *Guardian* newspaper, began publishing stories based on documents he got from Edward Snowden, a twenty-nine-year-old former government contractor who had lifted a trove of secret documents while working for the National Security Agency. Those documents detailed, among other things, an NSA surveillance program, known as PRISM, which captures data from Google, Facebook, YouTube, Skype, Apple, Microsoft, and a video-chat server in the Mideast known as Paltalk. The NSA's snooping into phone calls and Internet traffic is part of a broader global trend toward increased surveillance of individuals. In 2011 alone, cellular phone providers in the United States provided customer-calling data to law enforcement officials some 1.3 million times, and those law-enforcement requests don't always need to get a search warrant.[100] Each year, according to the *Economist*, South Korean authorities make more than 37 million requests to see communications data on its citizens. (The country has about 50 million people.) In the UK, police make about 500,000 such requests per year. (The UK has about 63 million people.) In India, the government is considering a plan that will rout all communications through its own servers, a system that could allow it to eavesdrop without telling Internet providers.[101]

In short, as we have become more connected, we have made it easier for our government to monitor nearly everything: where we go, who we talk to, who sends us e-mail, and what we search for on the Internet. And it's not just the government tracking us. So are advertisers and marketing companies.

In 2012, Alexis Madrigal, a writer for the *Atlantic* magazine, tracked all of the companies that were tracking him online. His conclusion: in one thirty-six-hour period, he was tracked by 105 different companies. Madrigal listed the outfits that followed his online clicks: "Acerno. Adara Media. Adblade. Adbrite. ADC Onion. Adchemy. ADiFY. Ad-Meld. Adtech. Aggregate Knowledge. AlmondNet. Aperture. AppNexus. Atlas. Audience Science. And that's just the As . . ." Madrigal continued, writing that advertisers and data-management companies are collecting the data so that they can "show you advertising that you're more likely to click on and products that you're more likely to purchase."[102]

While some people claim they are not worried about the government surveillance programs—"if you don't have anything to hide, then you have nothing to worry about," is their stock response—the potential for abuse of these surveillance operations is obvious, particularly when it comes to journalists who may be pursuing or publishing stories that the government doesn't like. Anti-government activists could also be targeted by government snooping.

All of this is worrisome and leaves people who are concerned about privacy with few choices. In a 2012 article in the *New York Times*, Matt Blaze, a professor of computer and information science at the University of Pennsylvania, said that when it comes to privacy, consumers have a choice: "Don't have a cellphone, or just accept that you're living in the Panopticon."*

---

* Designed in the late eighteenth century by Englishman Jeremy Bentham, the Panopticon ("all-seeing") was to be a round-the-clock surveillance machine. It would allow a watchman to observe all of the inhabitants of an institution without them being able to tell whether or not they were being watched. For more on the Panopticon, see: Cartome.org, http://cartome.org/panopticon1.htm.

Digital communications are fundamentally changing the way we live, learn, and travel. And yes, they come with some enormous downsides, including the loss of privacy.

But it's also abundantly clear that digital communications are fostering the exchange of ideas, and that exchange is further enabling liberty, freedom, and innovation. In his much-lauded 2011 book, *The Better Angels of Our Nature: Why Violence Has Declined*, cognitive scientist Steven Pinker writes that "successful innovators not only stand on the shoulders of giants; they engage in massive intellectual property theft, skimming ideas from a vast watershed of tributaries flowing their way . . . Societies that are marooned on islands or in impassable highlands tend to be technologically backward. And morally backward too."[103] Digital communications are allowing rivers of ideas to flow all over the world. And as those ideas flow, innovators are able to improve upon them.

Not every innovation is a net positive for human society; some create new problems. But it is undeniable that ongoing innovation is helping more people to live better lives than ever before.

I'll prove that point in the next chapter.

3

NEVER HAVE SO MANY
LIVED SO WELL

In 1971, an epidemiologist, Abdel R. Omran, wrote that from ancient times until about the mid-1600s, humans were "caught between the towering peaks of mortality from epidemics and other disasters and the high plateaus of mortality dictated by chronic malnutrition and endemic diseases." The result, he concluded, was that "life expectancy was short and human misery was assured."[1] Throughout much of recorded history, we humans have lived short, poverty-stricken lives. Epidemics and disease ran rampant and long life spans were rare.

That is no longer true. In 1900, the average US life span was about 47 years. Today, it's nearly 80.[2] People living in the world's poorest countries are living longer, too. In 1970, the average life span in the least-developed countries was 43 years. In 2011, it was 59 years.[3]

In nearly every country on the planet, disease and premature death are on the run. In 1990, according to the World Health Organization, 61 out of 1,000 babies would die by age one. By 2010, that number had dropped to 40.[4] Fewer women are dying during pregnancy and childbirth. Between 1990 and 2010, the number of women dying because of complications from pregnancy or birth was nearly halved, falling by some 3.1 percent per year over that time frame, according to the WHO.[5] There's positive news on the AIDS front. The disease hit its peak in the late 1990s, and by 2010 new infections were down by 20 percent when compared to 1997.[6]

There's also good news on the literacy front. In 1970, about 47 percent of all the adults on the planet were literate.[7] By 2009, according to the United Nations, the adult literacy rate was 83.7 percent.[8] However, some 775 million adults still cannot read or write, and of that number about 500 million of them are women.[9] Still, the progress is clear: as incomes continue rising, literacy rates will continue improving.[10]

Poverty is declining. In their 2013 book, *Conscious Capitalism*, John Mackey and Raj Sisodia point out that two hundred years ago, "85 percent of the world's population lived in extreme poverty (defined as less than \$1 per day); that number is now only 16 percent."[11] Data from the World Bank confirms that over the past couple of decades, poverty has been on a downward trend. Between 1990 and 2010, the percentage of people on the planet who are living in extreme poverty, which the World Bank defines as under \$1.25 per day, dropped by half. In 1990, about 43 percent of the people in the developing world were living in extreme poverty. By 2010, that figure had fallen to 21 percent.[12]

A 2009 study by two economists, Maxim Pinkovskiy from the Massachusetts Institute of Technology, and Xavier Sala-i-Martin from Columbia University, also found sharp declines in poverty. "Using the official \$1 per day line, we estimate that [from 1970 to 2006] world poverty rates have fallen by 80 percent." Their paper, "Parametric Estimations of the World Distribution of Income," which was produced for the National Bureau of Economic Research, found that over the past four decades or so, "measures of global welfare increased by somewhere between 128 percent and 145 percent."[13]

More people are living in freedom. According to Freedom House, in 1972, there were 44 countries that were considered "free." By 2011, that number had nearly doubled, to 87. Over that same time period, the number of countries that were classified as "not free" declined from 69 to 48.[14] Peter Wehner, an author, journalist, and fellow at the Ethics and Public Policy Center, has called it correctly: "In every corner of the globe, the tide is with human freedom and dignity."

In *The Better Angels of Our Nature*, Steven Pinker underscores and celebrates this remarkable progress. He demonstrates that violence is rapidly declining and that "we may be living in the most peaceful time in our species' existence."[15] Of the many reasons for this, Pinker points to the rise of cities, education of the masses, the rule of law, and the rise of global commerce. While he makes many valid points, it's also clear

Declining Global Poverty for Various Income Levels, 1970–2006

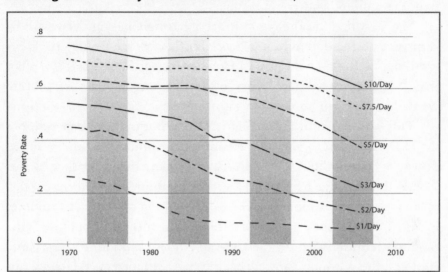

Source: Maxim Pinkovskiy and Xavier Sala-i-Martin.[16]

that things are improving because we are creating technologies that are raising living standards.

Those living standards are rising because we are finding and utilizing far more resources, from antimony to zinc, to produce more useful tools, from telephones and plows to online education systems and computers. By making all of them Smaller Faster Lighter Denser Cheaper, we are getting richer.

Some extreme environmentalists as well as some neo-Malthusians think that finding and using so many resources is not only bad news, but also an indicator of imminent ruination. All of the planet's resources are limited, the fearmongers remind us, and because they are limited, it must be true that we are running out of them—despite few indications that that will happen anytime soon. Remember the "peak oil" theorists, who got loads of media attention in the 1990s and 2000s? They are on the run. Hate the oil and gas industry if that's what makes you happy, but the history of that sector is one of remarkable ingenuity. Prices and technology are always combining to unlock hydrocarbons once thought unreachable. Put another way, the more oil and gas we find, the more oil and gas we find.

Many natural resources are, in fact, getting Cheaper. That will be surprising to the doomsayers who claim that we are facing "peak everything," as the title of Richard Heinberg's 2007 book put it.[17] Rather than "waking up to the century of declines" as the author's subtitle warned, we've seen remarkable gains.

This was shown in a 2010 analysis by John Boyce, an economist at the University of Calgary, who examined the production and prices of dozens of commodities over a time span of more than a century. Boyce looked at things like beryllium, mercury, uranium, molybdenum, aluminum, helium, diamonds, and gold from 1900 to 2007. His finding: for forty-eight of the eighty-one minerals, the real prices of those commodities fell even though the per-capita consumption of those same commodities was increasing. What did Boyce point to as the reason for those price reductions? Better technology. "The only way to get rising output with falling prices is for the supply curve to be shifting faster than the demand curve, which is generally associated with technological change," Boyce explained. In other words, we are getting better at finding and developing more resources.

To illustrate his claim, Boyce points to the Kern River oilfield in California, which was discovered in 1899. By 1942, the field had produced 278 million barrels of oil, and analysts believed that only some 54 million barrels of oil remained to be exploited. And yet, says Boyce, "by 1988, an additional 736 million barrels were produced and estimates of remaining reserves were raised to 970 million barrels." By 2010, Boyce said, after Kern River had produced a total of more than 2 billion barrels, still more oil was being found in the field. The latest estimates were predicting that about "627 million barrels remain recoverable" in the Kern River field.[18]

In 2011, the *Economist* published a graphic that showed the declines in the magazine's industrial commodity-price index between the mid-1800s and 2011. When measured in real-dollar terms, the cost of a basket of industrial commodities had declined by about 50 percent over a period covering more than 150 years.

Cheaper: The Trend in Industrial Commodities, 1850–2011

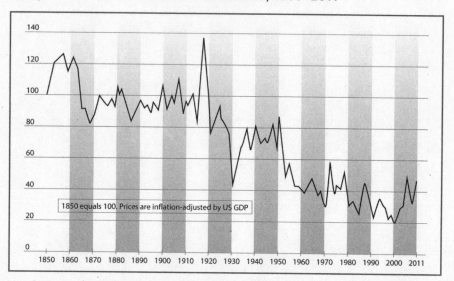

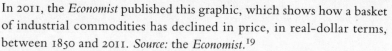
1850 equals 100. Prices are inflation-adjusted by US GDP

In 2011, the *Economist* published this graphic, which shows how a basket of industrial commodities has declined in price, in real-dollar terms, between 1850 and 2011. *Source:* the *Economist*.[19]

I'm not claiming that the Earth's resources are infinite or that we won't one day run out of some commodities or rare elements. What's clear is that we are continually figuring out ways to do more with less, to make things Smaller Faster Lighter Denser Cheaper, a trend that can easily be seen in manufactured goods.

In 1980, the average global cost of a solar photovoltaic module (which converts sunlight into electricity) was about $23 per watt.[20] By 2010, that price had fallen to about $2 per watt.[21] Back in 2005, I installed 3,200 watts of solar panels on the roof of my home in Austin, Texas. Since then, the price of photovoltaic panels has fallen by about half. Clearly, installing solar panels will still be expensive due to the high cost of labor as well as equipment like inverters and mounting brackets. But the cost-per-panel trends are positive. In 2013, First Solar, one of the biggest US-based producers of solar panels, claimed that it will be able to produce panels costing just $0.40 per watt by 2017.[22] If

Cheaper: The Trend in Photovoltaic Prices, 1980–2010

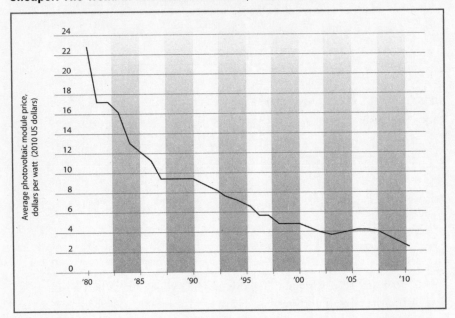

Global solar energy use is growing rapidly—up 58 percent in 2012 alone. That surge in use is due in large part to the availability of Cheaper photovoltaic panels that convert sunlight into electricity. And while it's good to be bullish on Cheaper solar, keep scale in mind. In 2012, solar's contribution to global energy demand was about 400,000 barrels per day of oil equivalent. Total global demand was roughly 250 million barrels of oil equivalent per day, or about 625 times as much.[23] *Source:* National Renewable Energy Laboratory.[24]

First Solar and other makers continue reducing costs, solar energy will be deployed more widely.

The pundits who warn of imminent resource depletion routinely discount the fact that we are wringing more and more value out of the energy that we consume. In 1970, American consumers drove about 1.1 trillion miles, and domestic airliners flew some 2 billion miles. That same year, US oil consumption was 14.7 million barrels per day. Forty years later, in 2010, Americans drove more than 2.9 trillion miles, and domestic airlines flew 5.9 billion miles. That year, the United States consumed an average of 19.1 million barrels of oil per day. Thus, over a period of four decades, Americans nearly tripled the number of miles they drove, and domestic airlines nearly tripled the number of miles flown; yet domestic oil consumption increased by just 30 percent over that time period.

How do we explain what happened? The companies that build airplanes and cars made operating their machines Cheaper by cutting their fuel use. In 1970, it took 10,185 Btu to move a single passenger one mile on an airliner. By 2008, that number had fallen to 2,931 Btu, a 71 percent reduction. A similar trend can be seen in passenger cars. In 1970, it took 4,842 Btu to move a single passenger one mile in a passenger car. By 2008, that figure had declined to 3,501 Btu, a reduction of about 28 percent.[25]

Travel has become commonplace because our transportation machines are Faster Lighter Cheaper than ever before. In 1903, Orville and Wilbur Wright made the first controlled, sustained flight with an airplane by flying at about 30 miles per hour (48 km/h).[26] Just 110 years later, Virgin Galactic, a space tourism company, tested a rocket ship that traveled at 1.2 times the speed of sound. (For reference, 1.2 Mach is 913 mph or 1,469 km/h.) The company hopes to begin offering rides into space, for $200,000 per person, in the near future.[27] While Joe Six Pack won't be traveling into space anytime soon, it's abundantly obvious that travel has become Faster Cheaper and safer.

In 1970, 52,627 Americans were killed in car accidents. By 2009, that number had fallen to 33,808. Put another way, the fatality rate per 100 million vehicle-miles traveled fell from 4.7 in 1970 to just 1.1 in 2009. The reduction in air-travel fatalities is even more striking. In 1970, the fatality rate per 100 million aircraft-miles traveled was 5.438. By 2009, that number was 0.688.[28] Thus, even though the total number of aircraft miles had nearly tripled, the fatality rate fell dramatically.

Even as we are traveling Faster Cheaper and safer than before, air quality has improved. In the United States, emissions of key air pollutants like sulfur dioxide and volatile organic compounds have been falling. In 1990, sulfur dioxide emissions were 23 million tons. In 2005, according to the Environmental Protection Agency, they were just 14.7 million tons.[29] In 1990, VOC emissions were 23 million tons. By 2005, they had fallen to 15 million tons.[30] Those numbers are remarkable enough by themselves. But they become even more astonishing when you consider that those dramatic reductions occurred over a period when US energy consumption increased by 19 percent.[31]

Thanks to better agricultural techniques, we are producing more food. The combination of hybrid seeds, better pest control, and more accurate application of fertilizers allows farmers to produce more food per acre of farmland than ever before. Between 1950 and 2010, global grain yields increased by an average of nearly 1.9 percent per year.[32] Those increasing yields cause more people in the developing world to give up subsistence farming and seek better opportunities in cities.

The growth of cities along with increasing availability of low-cost, high-speed Internet connectivity and satellite television are making the world Smaller. News and information are traveling Faster than ever before. The result: we live in a highly networked planet in which people living in repressed conditions are more able to see what life is like for those who have liberty.

For millennia, women were considered second-class citizens. Today, women are more free and better educated than ever before. At the 2012 Olympics, for the first time, all of the countries taking part in the Games included women on their teams.[33] Two women from Saudi Arabia, long one of the world's most repressive and backward regimes, were allowed to compete. With a handful of exceptions—Saudi Arabia is one—women are allowed to vote in nearly every country. A century ago, the situation was just the opposite, and women's suffrage was rare, with New Zealand, Finland, and Norway the notable exceptions.[34] In US colleges, women are no longer rare. Instead, they are the majority. In 2012, 54 percent of all first-time, full-time freshman students entering college were women.[35]

The facts are simply indisputable: never have so many lived so well, or so free. Yet despite this astounding progress, there remains an entrenched and powerful interest group that believes we humans are doing too much, that we must reduce our consumption of everything, return to our agrarian past and employ what one prominent catastrophist calls "a new civilizational paradigm."

Following such a path would be disastrous. Just as we are accelerating the trend toward Smaller Faster Lighter Denser Cheaper in nearly every sector, some people want to turn back the clock and wrench defeat from the grasping fingertips of victory. And that poses the essential question: will we continue innovating, embracing technology, and getting richer, or will we listen to those who are advocating degrowth?

4

BACK TO THE PAST

THE PUSH FOR "DEGROWTH"

In September 2011, Ted Nordhaus and Michael Shellenberger, the founders of the Oakland-based center-left think tank Breakthrough Institute, wrote an essay for *Orion* magazine in which they coined a phrase that neatly sums up the worldview of many environmentalists and environmental groups. Nordhaus and Shellenberger called this view "nihilistic ecotheology."

That worldview, they said, comprises "apocalyptic fears of ecological collapse, disenchanting notions of living in a fallen world, and the growing conviction that some kind of collective sacrifice is needed to avoid the end of the world." The eco-nihilists have "nostalgic visions of a transcendent future in which humans might, once again, live in harmony with nature through a return to small-scale agriculture, or even to hunter-gatherer life."[1]

Other analysts have described this same worldview. In 2013, David Deming, a geologist and professor at the University of Oklahoma published an op-ed in the *Wall Street Journal* that declared that "modern environmentalism is based on emotionalism . . . and the myth of primitive harmony."[2]

There are plenty of examples from the Green Left that reflect the worldview described by Nordhaus, Shellenberger, and Deming.

The myth of primitive harmony that pervades much of modern environmentalism has deep historical roots in Western culture. It can be traced back to Rousseau in the eighteenth century, to Thoreau in the nineteenth century, to Edward Abbey in the twentieth century, and is readily apparent in the present-day rhetoric of Greenpeace, the

Sierra Club, and numerous other environmental groups and leaders. Many modern advocates for conservation and preservation see a world swirling toward disaster and the need for drastic action that would curtail nearly every activity of modern life in order to save the planet, and presumably, ourselves.

They contend that we ignore their warning at our peril. As the French writer Pascal Bruckner explains, the catastrophists alone "see the future clearly while others vegetate in the darkness." These predictors of apocalypse, says Bruckner, are trying "desperately to awaken us, to convince us of planetary chaos." Or as the advertisements for Al Gore's film, *An Inconvenient Truth,* warned: "Humanity is sitting on a time bomb." Only by taking dramatic action—forsaking hydrocarbons for renewable energy—will humanity be able to avoid a "tail-spin of epic destruction."[3] Since the release of his film, Gore has frequently referred to what he calls the "climate crisis."[4] He's also said we are facing a "planetary emergency."[5]

Averting the looming (pick your favorite term) catastrophe, time bomb, crisis, or emergency, requires us to hew to their worldview, one in which we humans are the problem and the Earth is the object to be saved. The biggest and most influential environmental groups routinely preach a message of doom. They regularly claim, for instance, that technology is dangerous (their opposition to nuclear and GMOs are obvious examples of this mindset) and that industrial development must be stopped in order to the save the planet. However, the painful paradox is that they are aiming to stop many of the innovations that are helping to improve the environment and raise the living standards of millions of people. They are also promoting energy policies that would be ruinous for the environment they say they want to protect.

Let's start with Greenpeace, one of the world's biggest environmental groups, which has an annual operating budget of more than $300 million.[6] Throughout its history, Greenpeace has been stridently antinuclear. It is also opposed to the use of hydrocarbons. Rex Weyler, a founder of Greenpeace International, has been among the leaders of the degrowth movement, an effort to stop economic growth in order to—in theory—save the planet.

In 2011, Weyler published an article on Greenpeace's Web site that said, "Degrowth is an important, natural concept that our society needs to understand . . . As we learn to share and live modestly, our ecosystems can recover and provide us with nature's bounty. The best way for poor nations to avoid deeper poverty is to protect their ecosystems from plunder." He adds that the degrowth movement "advocates richer, more rewarding lives with less material stuff. Our economic efforts should focus on providing basic needs to everyone in the human family, rather than enriching a few, while others starve."[7]

Weyler's agenda is similar to what Naomi Klein promoted in a 2011 cover story in the *Nation,* a magazine that has long been the vanguard for the American Left. In "Capitalism vs. the Climate," Klein declared that we humans "have pushed nature beyond its limits" and therefore need "a new civilizational paradigm, one grounded not in dominance over nature but in respect for natural cycles of renewal." Klein went on to claim that the solution for saving humankind from global warming "requires that we break every rule in the free-market playbook and that we do so with great urgency."

Klein, who has authored several books, including *The Shock Doctrine: The Rise of Disaster Capitalism*, wrote that the solution is to "scale back overconsumption" and "heavily regulate and tax corporations, maybe even nationalize some of them" while recognizing "our debts to the global South."[8] Let's ignore for a moment the socialist—and even communist—implications of Klein's essay and instead focus on how she plans to save Mother Earth: The "real climate solutions" she claims, are projects that are controlled at "the community level, whether through community-controlled renewable energy, local organic agriculture or transit systems genuinely accountable to their users."

In 2012, the Worldwatch Institute, a left-leaning environmental research group based in Washington, DC, said it was time for governments to "tax ecologically harmful industries" that make "unhealthy or unsustainable products." The money from the levy "could curb the growth of harmful industries and of the worst forms of consumption, while also raising government revenues to build green infrastructure—such as improved water and sanitation systems, public transit, renewable energy, and bicycle lanes."[9] The Worldwatch Institute's Erik Assadourian explained that a primary goal of the degrowth movement was to shrink the economies of what Worldwatch considers "overdeveloped" countries. In addition, the group wants to create "a steady-state economic system that is in balance with Earth's limits . . . and restore the planet's ecological systems."[10]

Bill McKibben, a man the *New York Times* has called "perhaps the nation's most effective grass-roots environmental advocate," is also a proponent of degrowth.[11] McKibben has written that "our systems and economies have gotten too large . . . we need to start building them back down. What we need is a new trajectory, toward the smaller and more local."[12] McKibben is a prolific author as well as the founder of 350.org, an organization that aims to drastically cut global carbon dioxide emissions. In a 2012 essay for *Rolling Stone* magazine, McKibben was even more blunt about his antibusiness stance. The global warming issue, he wrote, is "not an engineering problem . . . it's a greed problem."[13]

In 2013, Matthew C. Nisbet, a communications professor at American University, wrote a long profile of McKibben, in which he pointed to McKibben's "roots in the deep ecology movement." McKibben's goal, Nisbet says, has been to "generate a mass consciousness in support of limiting economic growth and consumption with the hope of shifting the United States toward localized economies, food systems, and 'soft' energy sources."[14]

McKibben justifies his push for degrowth by claiming that we are on the precipice of disaster. In 2013, he told the *Atlantic* magazine, "In a sense, the world as we knew it is already over. We have heated the Earth, melted the Arctic, and turned seawater 30 percent more acidic." He continued, saying, "The only question left is how much more fossil fuel we'll burn, and hence how unfamiliar and inhospitable we'll make our home planet."[15]

McKibben and his fellow travelers believe that salvation lies in pursuing low density in both energy production and food production. But the precise opposite is true. Density is green. It's only by increasing the density of our energy and food production that we will be able to meet the demands of our growing population. And yet, the Sierra Club,

Greenpeace, and many other groups want to pave the world with low-density wind turbines. Not only do they insist on renewable energy; they want us all to live on homesteads equipped with a be-draggled organic garden, a compost pile, and maybe a few scrawny chickens. And of course, there's no Volvo station wagon in the drive-way, only a pair of battered 3-speed Sturmey Archer bicycles, and one of them has a flat.

There's no doubt that this return-to-nature idea has some appeal. Clearly, we've paved parts of paradise and lost beautiful places to de-velopment and ruined others with pollution. We need to protect our wild places. I'm a birdwatcher, a beekeeper, an active hiker, and paddler. On a 2013 trip to the Galapagos Islands, I was gobsmacked by the stark beauty of the islands, by the variety and loveliness of Darwin's finches, by the playfulness and friendliness of the sea lions, the enormity of the whales, and the clarity of the islands' air and water. The Ecuadorian government has done a wonderful job of protecting the islands even as the tourism industry has grown dramatically. Wild places fortify us. Urban parks and open space allow us to exercise and socialize. It's be-yond debate that we need clean air, drinkable water, wholesome food, and livable cities.

But our salvation cannot be found in returning to the 40-acres-and-a-mule-*Green-Acres* plan put forward by McKibben, Klein, and their allies. Nor can it be achieved by imposing degrowth—and all of the unemployment and wrenching poverty that would surely come with such a scheme. Instead, our future depends on embracing technology.

It's only through a pro-business, pro-innovation, and pro-human outlook that we will be able to succeed. It's only by creating wealth that we will be able to support the scientists, tinkerers and entrepre-neurs who will come up with the new technologies we need. It's only by getting richer that we will be able to afford the adaptive measures we may need to take in the decades ahead as we adjust to the Earth's ever-changing climate. It's only by using more energy, not less, that we will be able to provide more clean water and better sanitation to the poorest of the poor. It's only by accepting the inevitability of what

Roger Pielke Jr. of the University of Colorado calls a "high-energy planet" that impoverished countries like India, Vietnam, Thailand, Nigeria, South Africa, Uganda, and others will be able to bring their people into the modern world. Put short, we need more innovation, not less. We need new tools and techniques. We need to fan the flames of Smaller Faster Lighter Denser Cheaper.

Unfortunately, the vanguard of the Green Left continues to promote an antibusiness, anti-innovation, antimodern energy, and in some cases, an anti-human outlook. The Green Left's romanticization of the past, along with its continuing claim that renewable energy and organic agriculture are the only way forward, ignores the deprivation, lack of social, intellectual, and economic mobility, and short life spans that dominated preindustrial societies.

Bill McKibben's Energy-Starvation Plan

Bill McKibben is on a quest to stabilize the concentration of carbon dioxide in the Earth's atmosphere at 350 parts per million, a level that he and some others claim is the ideal. McKibben, the author and environmental activist, is fond of saying "do the math." Okay. Let's.

The arithmetic is laid out in McKibben's 2010 book, *Eaarth: Making a Life on a Tough New Planet,* in which he said that if humans want to "stabilize the planet" and reduce the atmospheric concentration of carbon dioxide to 350 parts per million, then "we need to cut our fossil fuel use by a factor of twenty over the next few decades."[16] Let's consider what McKibben's twentyfold reduction in hydrocarbon use might look like.

In 2012, global use of hydrocarbons—coal, oil, and natural gas—averaged about 218 million barrels of oil equivalent per day.[17] Reducing that by a factor of 20 would take global

hydrocarbon use down to about 11 million barrels of oil equivalent per day, which is roughly the total amount of energy used
by India in 2012.[18] (For an alternative comparison, consider that
in 2010, global gasoline consumption averaged about 22 million
barrels per day.)[19]

To make doing the math easier, let's convert those 11 million barrels of oil equivalent per day into liters. The volume of
a barrel is 159 liters. Therefore, under McKibben's plan for a
twentyfold cut in hydrocarbon use, the daily ration of hydrocarbons for the entire population of the planet would be 1.75
billion liters of oil equivalent. If we were able to divide that
amount of energy equally among seven billion people—not an
easy task—you end up with a ration of 0.25 liters (25 centiliters)
of oil equivalent per person, per day.

Let's put that into perspective. In 2011, the average resident of planet Earth consumed about 4.9 liters (1.3 gallons) of
oil-equivalent energy per day from hydrocarbons.[20] Therefore,
if McKibben's plan were enacted, each of the seven billion
residents of the planet would be allowed a daily ration of hydrocarbons that *wouldn't fill an average-size soda can.* And keep in
mind all of these calculations assume absolutely no population
growth over the coming decades and that the world's population will stay at seven billion.

Let's do the math by considering what McKibben's energy-
starvation diet might mean for some of the world's poorest people,
and further assume that they are allotted the same 25 centiliters
of oil equivalent per person per day as every other person on the
planet. In 2011, the average resident of Bangladesh used the energy
equivalent of about half a liter of oil per day.

Under McKibben's prescription, then, the average Bangladeshi would be required to cut his/her energy use by about
half. McKibben's meager diet of hydrocarbons would be equally
ruinous for residents of India, a country in which about four

hundred million people live without electricity. In 2011, the average Indian consumed about 1.5 liters of oil equivalent per day and of that, about 1.3 liters came from hydrocarbons.[21] Thus, under the McKibben plan, each Indian would have to reduce his/her energy consumption by a factor of five. Or consider China, where the average resident consumed 6.2 liters of oil equivalent per day, of which about 5.5 liters came from hydrocarbons. Meeting McKibben's goal would require the average Chinese citizen to cut hydrocarbon use by a factor of 22.

We can also do the math by ignoring the rest of the world and focusing solely on the United States. Current hydrocarbon use in America is about 20 liters per capita per day. Cutting that by a factor of 20 would mean each American could use just 1 liter of hydrocarbons per day. That amount is about half of the current per-capita hydrocarbon consumption in Peru.[22]

Yes, these numbers are rather hard to digest. So here's one more comparison that might drive—pun intended—the point home. On a ration of 1 liter of gasoline per day, a Prius driver in the United States would be limited to no more than 13 miles of driving per day. The driver of a Chevrolet Suburban would be allotted about 4 miles per day.[23]

And don't plan on doing any driving with an electric car, either. McKibben isn't just opposed to hydrocarbons. He's also antinuclear. In 2012, McKibben dismissed the potential of nuclear energy, declaring that "It's too expensive. It's like burning $20 bills to generate electricity." He quickly added: "The good news is we are getting really a lot better at using the soft renewables like sun and wind."[24]

Alas, once again, McKibben doesn't do the math. In 2012, solar and wind energy provided slightly more than 1 percent of all global energy consumption. Together, they provided about 2.8 million barrels of oil equivalent per day while all global energy needs totaled 250 million barrels of oil equivalent per

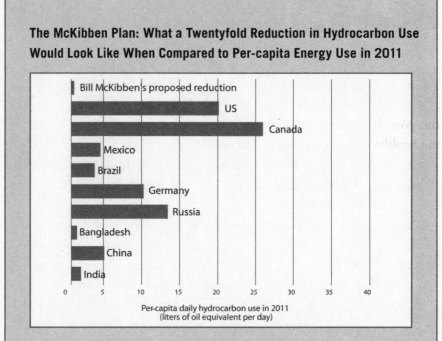

The McKibben Plan: What a Twentyfold Reduction in Hydrocarbon Use Would Look Like When Compared to Per-capita Energy Use in 2011

Per-capita daily hydrocarbon use in 2011
(liters of oil equivalent per day)

Sources: BP Statistical Review of World Energy 2013; Bill McKibben, *Eaarth: Making a Life on a Tough New Planet* (Toronto: Vintage, 2010).

day.[25] As I will demonstrate in a later chapter, the critical problem with renewable energy in general—and wind energy and biofuels in particular—is their low power density. Wind and biofuels simply require too much land to be viable on a large scale because land dedicated to renewable energy production cannot be used for housing, food production, or for parkland. Therefore, McKibben's claim about "soft renewables" getting "a lot better" is little more than spin.

The bottom line here is obvious: if the countries of the world decided to embrace McKibben's antinuclear, antihydrocarbon proposals, the result would be dire poverty for billions of people around the world. McKibben may couch his rhetoric in environmental terms, but his proposals are a prescription for economic suicide.

In 2002, two anthropologists, Richard H. Steckel of Ohio State University, and Jerome C. Rose from the University of Arkansas, published a study based on their analysis of more than 12,500 Native American skeletons that date from the pre-Columbian era. Steckel and Rose found that few people who lived in hunter-gatherer societies in the Americas survived past the age of fifty. And, as John Noble Wilford of the *New York Times* explained, Steckel and Rose determined that "in the healthiest cultures in the 1,000 years before Columbus, a life span of no more than 35 years might be usual."[26] Yes, the Cherokees, Choctaws, Seminoles, and Apaches may have been noble, but they sure didn't live long. Even shorter life spans—about 18 years—were common among the ancient Greeks. Among the Romans, it was about 22 years.

The Greeks, Romans, Zunis, and Navajos of yesteryear ate organic food, or they didn't eat at all. Renewable energy? They had no other choice. The sun provided the energy for the biomass—wood, dung, straw—that they used for their heating needs. The wind provided motive power for their boats. And yet, despite their all-organic, all-renewable diets, nearly all of them were dead by age thirty-five, with no opportunities to even have a mid-life crisis. No motorcycle, Porsche, or mistress for those Ultra-Greens of centuries past, nor even the prospect of them.

Despite the short and brutish living conditions that dominated human settlements for millennia, the myths about primitive harmony are abundant in the literature. The myth is as old as the Garden of Eden, the place described in the book of Genesis where we humans allegedly first fell from grace. In addition to the Bible, the notion of primitive harmony can be seen in medieval times in the *Romance of the Rose*, written in the thirteenth century by the French poet Jean de Meun. The lengthy poem describes a world in which humans have simple tastes and the earth provides everything that is needed in abundance. People lived in harmony until their paradise was spoiled by the desire for money, power, and property.[27]

But it was the Swiss-born philosopher Jean-Jacques Rousseau who has likely done the most to popularize the notion of primitive harmony. In the 1700s, Rousseau came up with the idea of a "natural man" who was unspoiled by civilization.[28] In what may be his most famous work, *Discourse Upon the Origin and the Foundation of the Inequality Among Mankind* (written in 1754, published 1755), the philosopher wrote, "As long as men remained satisfied with their rustic cabins; as long as they confined themselves to the use of clothes made of the skins of other animals . . . as long as they undertook such works only as a single person could finish, and stuck to such arts as did not require the joint endeavors of several hands, they lived free, healthy, honest and happy."[29]

But as soon as "one man began to stand in need of another's assistance" claimed Rousseau, "all equality vanished" and "slavery and misery were soon seen to sprout out and grow."[30]

Exactly 100 years after Rousseau wrote his *Discourse*, and fifty-six years after Thomas Malthus offered his grim view of the future, Henry David Thoreau published *Walden, or Life in the Woods,* which told of his two-year stint living in a modest cabin on the shore of Walden Pond in Concord, Massachusetts. Thoreau admonished readers to seek "simplicity, simplicity, simplicity." Thoreau himself had worked in his family's business, Thoreau and Company, which was renowned for the quality of its pencils. He'd invented a machine that made fine graphite for use in pencils.[31] And yet, Thoreau was antitechnology. He extolled the benefits of walking, and even though railroads were making travel Faster and Cheaper than ever before, he declared, "We do not ride on the railroad; it rides upon us."[32]

In the conclusion to *Walden*, Thoreau said that Americans "and moderns generally, are intellectual dwarfs compared with the ancients." He advised readers to "cultivate poverty like a garden herb . . . Do not trouble yourself much to get new things, whether clothes or friends . . . Sell your clothes and keep your thoughts. God will see that you do not want society."[33] (In 2004, John Updike wrote that Thoreau's *Walden* has "become such a totem of the back-to-nature, preservationist, anti-business, civil-disobedience mindset, and Thoreau so vivid a protester, so perfect a crank and hermit saint, that the book risks being as revered and unread as the Bible.")[34]

A little more than a century after Thoreau published *Walden*, Rachel Carson published one of the most famous books of the Green canon: *Silent Spring.** And like *Walden*, it lauds the innocence and purity of the past. Carson's 1962 book claims that "there was once a town in the heart of America where all life seemed to live in harmony with its surroundings."[35]

In 1972, the Club of Rome published *Limits to Growth,* a report that predicted widespread calamities due to "accelerating industrialization, rapid population growth, widespread malnutrition, depletion of non-renewable resources and a deteriorating environment." If those trends continue, the report said, "the most probable result will be a rather sudden and uncontrollable decline in both population and industrial capacity."[36]

Edward Abbey, one of the patron saints of American environmentalism, didn't soft-pedal his misanthropy. In his 1971 book, *Beyond the Wall,* Abbey wrote: "We humans swarm over the planet like a plague of locusts, multiplying and devouring. There is no justice, sense or decency in this mindless global breeding spree, this obscene anthropoid fecundity, this industrialized mass production of babies and bodies, ever more bodies and babies." Abbey declared that we must "learn to control, limit and gradually reduce our human numbers . . . To aid and abet in the destruction of a single species or in the extermination of a single tribe is to commit a crime against God, a mortal sin against Mother Nature."[37]

Abbey's 1971 essay repeats a misanthropic view that has been around for decades. It was published just three years after Paul Ehrlich published *The Population Bomb,* a book that was commissioned and published by the Sierra Club. In Ehrlich's telling, population control was absolutely essential to avoid catastrophe. "Conscious regulation of human numbers must be achieved," wrote Ehrlich.[38] In a foreword to the book, David Brower, who was then the head of the Sierra Club, wrote that

---

* *Silent Spring* was largely responsible for the US-imposed ban on the pesticide DDT, which had proven effective at killing mosquitoes, which spread malaria. But DDT also had many negative consequences on wildlife and in particular on birds of prey. The ban on the insecticide has been credited as a significant factor in the comeback of the bald eagle and other raptors. The ban has also hampered the fight against malaria.

environmental groups "have been much too calm about the ultimate threat to mankind."[39] Ehrlich's book sold more than two million copies. But Ehrlich was merely repeating an idea that was expressed starkly in a 1955 essay published in *Science* magazine, which famously declared, "The world has cancer and the cancer is Man."[40]

That view—that we humans are a cancer—lives on among some of the world's highest-profile environmentalists. In 2013, David Attenborough, the British naturalist and filmmaker who has gained fame for his many documentaries, declared that humans are "a plague on the Earth."[41]

If humans are a plague—of locusts or something else altogether—then the obvious solution for many on the Green Left is that humans must be stopped. More particularly, their economies and businesses must be stopped. As Bolivia's socialist president, Evo Morales, declared in 2009, "Either capitalism dies, or Mother Earth dies."[42]

In 2009, the British author and journalist George Monbiot, who writes a regular column on environmental issues for the *Guardian*, averred that "an ordered and structured downsizing of the global economy" is what is needed. He also said that the planet can only support some two billion people, and even that number "is surely the optimistic extreme."[43]

The idea of downsizing the economy and limiting growth—which ultimately means limiting business and innovation as well as human population—lies near the heart of prescriptions put forward by some proponents of the "planetary boundaries" theory, which is really just a new name for the concepts put forward in *Limits to Growth* back in the 1970s, a report that was itself just another bit of rehashed Malthusianism. And yet the neo-Malthusian mindset endures. In 2011, three analysts, Will Steffen, Johan Rockström, and Robert Costanza, published a report in which they claimed to have identified specific boundaries for the planet—on issues like climate change, land use, water use, ozone depletion, and others—"beyond which humanity should not go." The three believe that the Earth has "intrinsic, nonnegotiable limits" and that we need to begin implementing moves that will allow us to live within those limits. But it's the implementation part of their prescription that creates the rub. They write:

Ultimately, there will need to be an institution (or institutions) operating, with authority, above the level of individual countries to ensure
that the planetary boundaries are respected. In effect, such an institution, acting on behalf of humanity as a whole, would be the ultimate
arbiter of the myriad trade-offs that need to be managed as nations and
groups of people jockey for economic and social advantage. It would,
in essence, become the global referee on the planetary playing field.[44]

The concept of a "global referee"—call him Big Brother for
environmentalists—who has the power to allocate resources on a whim,
is plenty scary. Nevertheless, the view that economic growth is bad also
pervades the thinking of the Sierra Club. The club, one of America's
oldest and most influential environmental groups, has opposed nuclear
energy since 1974 and says it will remain opposed to nuclear, pending
"development of adequate national and global policies to curb energy
overuse and unnecessary economic growth."[45]

It's unclear what "unnecessary economic growth" might look like
to someone who's unemployed. It's also unclear what "energy overuse"
looks like to a family living without electricity. What is clear is this: the
Sierra Club's opposition to economic growth—and therefore, energy
consumption, employment, and human development—stands in stark
contrast to what the people of the planet need right now.

Economic growth is essential if we are to have enough tax dollars
to fund our schools and universities, which have long been incubators
of innovation. Economic growth allows governments to have more
revenue, which can be used to support research in health care, energy,
and other sectors. Economic growth means more employment, which
leads to more optimism about the future. That optimism, in turn, en-
courages investment in new technologies.

The alternative is pessimism. Believing in degrowth means believing in poverty. Believing in degrowth means rejecting technology. It's
time to move past Ehrlich, the Sierra Club, McKibben, Klein, Greenpeace, and the rest of the neo-Malthusians. At the risk of depleting my
quota of hyphens, it's clear that we need more anti-neo-Malthusians. It's
only by fostering innovation—and the business and research investments

that drive it—that we can bring more people out of poverty. It's time to discard the romantic view of primitive harmony once and for all.

The next section looks at the modern-day innovators who are pursuing Smaller Faster Lighter Denser Cheaper. It also tells the story of the individual innovators and companies from years past who brought us Cheaper computers and transportation. It's about lasers and engines, drill rigs and batteries, cities and farms. The unifying theme for Part II is the never-ending push for Smaller Faster. Indeed, we are already living in a world in which distance can be measured in angstroms and time can be measured in attoseconds.

PART II:

Our Attosecond World

How We Got Here, Where We're Going,
and the Companies Leading the Way

5

ANGSTROMS AND ATTOSECONDS

In the 1870s, after years of vigorous debate, the British photographer Eadweard Muybridge settled a debate that had been raging for years. Using a bank of high-speed cameras, Muybridge took a series of photographs that proved once and for all that galloping horses do, in fact, take all of their hooves off of the ground at once. The secret to Muybridge's images: cameras with shutter speeds of one-thousandth of a second.[1]

Now fast forward—and I do mean *fast* forward—to the work being done by Canadian physicist Paul Corkum. Corkum's imaging machines are a quadrillion times Faster (that's 10^{15} times) than the ones used by Muybridge. Corkum is taking pictures of electrons, and in doing so he's measuring distances in angstroms and time in attoseconds.

Those two terms are not part of our regular vocabulary, so let me pause for a microsecond to explain. By using specially tuned lasers that produce enormously high power densities, Corkum and his colleagues are peering into the inner workings of atoms, a region that is commonly measured in angstroms. (As noted earlier, an angstrom is one-tenth of a nanometer, or 0.1 billionth, 10^{-10}, of a meter.) Corkum and his colleagues are able to peer into the angstrom level by using laser pulses that are measured in attoseconds. An attosecond is a billionth of a billionth (10^{-18}) of a second, or a nano-nanosecond.

Corkum's work exemplifies our quest for Smaller Faster Lighter Denser Cheaper. "For centuries science has worked to measure Faster and Faster phenomena," Corkum said when I visited him at his office at the National Research Council in Ottawa, a beautiful 1930s-era

May 6, 2013: Paul Corkum stands inside the Joint Attosecond Science Laboratory in Ottawa, a facility shared by the University of Ottawa and the National Research Council Canada. The instruments in front of him are part of a suite of machines that allow Corkum and his colleagues to produce some of the world's shortest bursts of light. By mid-2013, the JasLab was generating laser pulses that were about 140 attoseconds. One attosecond is 10^{-18} second, or a billionth of a billionth of a second. To imagine what an attosecond is, Corkum explains that "comparing one attosecond to one second is like comparing one second to the age of the universe."[2] *Source:* Photo by author.

building that sits on the banks of the Ottawa River. The view from his office looked north across the river toward Quebec.

"Muybridge looked at horses. That was an important issue of his time," said Corkum, an affable man in his midsixties. Today, he said, there are many things that move fast that we want to understand better. A prime example, he continued, are the power flows in the circuits in a semiconductor. But what we really want to understand, he continued, is what happens inside atoms and molecules. In the quest to do that, Corkum is using lasers that concentrate a beam of light that is just

In the 1870s, these images, taken by British photographer Eadweard Muybridge, settled one of the big scientific debates of the day. By using cameras with shutter speeds of about 1/1000th of a second, Muybridge proved that sprinting horses take all of their feet off the ground at one time. Today, laser flashes can be measured in attoseconds, or 10^{-18} second, or 1/1000000000000000000 of a second. *Source*: Library of Congress, LC-DIG-ppmsca-23778.

100 microns wide, about the thickness of a human hair. (One micron is one-millionth of a meter or 10^{-6} meter.) The beam is then focused through a series of mirrors and concentrated onto a molecule. The intense burst of energy from the laser momentarily releases an electron that then acts, in effect, like a super-high-speed flash on a camera.

Corkum and his colleagues use techniques known as high-harmonic interferometry and high-harmonic spectroscopy. By matching their laser's wavelength to that of the distances between the objects in a molecule (the electrons, protons, etc.), Corkum and others who work in attosecond science are able to illuminate and measure those interstitial spaces.

For those who are not overly savvy on tomography, ionization, and quantum mechanics, Corkum offered this analogy: imagine that the attosecond pulses of the laser are strobe lights in a discotheque. In his laboratory, those laser light pulses are aimed at atoms and molecules instead of dancers on the disco floor. The result from both the disco and the laser pulse is the same: a freeze-frame image.

Corkum sees a multitude of possibilities that could emerge from his work. "Imagine if we could see what happens inside the molecule during chemical reactions." But Corkum got fully animated when he began discussing the evolution of lasers and the potential to create a 3-D map of an individual cell, a map that could show the location of every molecule in the cell. That might be possible, he said, because lasers are allowing scientists to manipulate light in ever-finer ways.*

Today, we carry laser pointers in our pockets and rely on laser readers to play the music on our CDs. But much of the work on lasers has focused on making them more powerful. And that brings us back to the continuing desire for Denser, or to be more specific, our push for power density. The areal power density of the lasers in Corkum's laboratory is 10^{14} watts per square centimeter.[3] To be certain, the laser beam that Corkum is using doesn't cover an entire square centimeter. Nevertheless, if we translate that areal power density into watts per square meter—the metric that I use for other calculations in this book— the numbers are astonishing. Doing so shows that the lasers in Corkum's basement-level laboratory have a power density of 1 exawatt (10^{18}) per square meter. For comparison, the power density inside the core of a nuclear reactor is about 300 megawatts (mega is the SI designation for million, or 10^6) or 300 million watts per square meter. Put another way, the power density of the lasers Corkum is using is about a billion (10^9) times more powerful than the power flows that are found inside the core of a nuclear reactor.

As Corkum and I discussed what practical technologies might come of the work being done in attosecond science, he reminded me that James Watson and Francis Crick were able to describe the double-helix structure of DNA thanks to scientists' ability to manipulate different parts of the light spectrum. Watson and Crick built on the work done by the British biophysicist Rosalind Franklin, who was using X-ray diffraction to probe the structure of DNA.[4] As you may recall from your high-school physics class, X-rays have shorter wavelengths than visible light.[5] In 1962, Watson, Crick, and another man, Maurice Wilkins,

---

* The laser—short for light amplification by the stimulated emission of radiation, an idea that stemmed from a paper that Albert Einstein published in 1905 on the idea of the photon—has undergone rapid improvement since the first working laser was developed in 1960. For more, see: University of Chicago Press, "The First Laser," undated, http://www.press.uchicago.edu/Misc/Chicago/284158_townes.html.

were awarded the Nobel Prize for their work.[6] Franklin, who died in 1958, was not recognized for her work.

Thanks to our ability to manipulate the very fast (light waves), scientists like Watson, Crick, Wilkins, Franklin, and Corkum have been able to unlock the secrets of the very small. We have a hunger for both Smaller and Faster. And the desire for Faster has been turbocharging innovation for centuries.

6

HOW OUR QUEST FOR FASTER DRIVES INNOVATION

We have a need for speed.

Whether it's the time needed to run a marathon, drive to Las Vegas, do the laundry, download a copy of "Heartbreak Hotel" from the iTunes store, or cross the Atlantic Ocean, we are obsessed with Faster.

Guinness World Records has a big section on speed, including "fastest 100 meters with a can balanced on head (dog)," and "fastest 100 meter hurdles wearing swim fins (individual, male)."[1] For the latter, the record is 14.82 seconds, which was set in 2008 by a German sprinter—or was he a swimmer?—named Christopher Irmscher.[2] Oh, and there's the "fastest time to hula hoop 10 kilometers (male)," a title held by an American, Ashrita Furman, who swiveled and shimmied the required distance in 1:25:09 in 2006.[3]

The quest for speed pervades everything we do, and it has been ongoing for millennia. The oldest known wheel was discovered in Mesopotamia and dates back to about 3500 B.C. Another 1,500 years or so would pass before the Egyptians managed to develop a modern wheel, that is, one with spokes.[4] How fast would those ancient Egyptian wheels rotate? If they were attached to a chariot traveling at about 50 kilometers (31 miles) per hour, they were likely turning at about 400 revolutions per minute.[5] Today, the impeller inside engine turbochargers rotate at speeds of up to 250,000 revolutions per minute, or about 6,000 times as fast as those Ben-Hur-era chariot wheels.

Our desire for Faster sea travel led to improvements in sails, hull design, and navigation techniques. The quest for Faster cars and airplanes spurred improvements in metallurgy, engine design, and aerodynamics.

Race car driver Bob Burman, about 1910. Burman set a number of records, including a land-speed record of 141.7 miles per hour on April 23, 1911. In 1916, he was killed in a crash while competing in a road race in California. He was thirty-one.[6] *Source*: Library of Congress, LC-DIG-ggbain-09237.

The need for Faster Internet connectivity has led to breathtaking advances in wired and wireless communications systems. Since 2000 or so, the speed of the average US home Internet connection has increased by some 900,000 percent.[7]

Roman cargo ships plying the Mediterranean moved at about 7 kilometers per hour or 4.5 miles per hour.[8] Today, a Boeing 737, the most popular jetliner ever built, travels at about 0.785 Mach, which is 518 miles per hour (833 kilometers per hour).[9] Thus, the Romans of today can easily travel more than 100 times as fast as their ancient cousins, and by jumping on an airliner, they can travel to nearly any place on the planet and be there within a day or two.

Columbus's first voyage across the Atlantic to the New World in 1492 took more than two months.[10] That famous trip launched a centuries-long effort to decrease the amount of time needed to get from Europe to America and vice versa. By the 1700s, sailing ships still needed six weeks or more to make the crossing. The never-ending push for Faster led to the steam engine. By 1845, the *SS Great Britain*, a steam-powered ship designed by the engineering genius Isambard Kingdom Brunel, was crossing the Atlantic in just fourteen days.[11] A bit more than a century later, in 1952, the ocean liner *SS United States*, designed by William Francis Gibbs, was making the same voyage in just three and a half days, a record that stands to this day.[12] But the *United States*, like other luxury ocean liners, were destined to go the way of the buggy whip. In the late 1950s and early 1960s, jetliners began traversing the Atlantic in a matter of hours.

Not only do we want machines that go Faster, *we* want to go Faster. We dream of sprinting like Usain Bolt, hurdling like Edwin Moses, and high-jumping (or in his case, high-flopping) like Dick Fosbury.[13] For as long as humans have been walking on two legs, they have been lining up and drawing lines in the dirt to see who can cover a given distance—wearing swim fins or nothing at all—in the shortest amount of time. Among the most popular events of the ancient Olympic Games, which date back to about the sixth century B.C., were foot races in which the runners competed in the buff. The races were of varying lengths and were measured in *stade,* which was 192 meters, or the length of the stadium.[14]

The obsession with Faster—and all the glory and cash that comes with it—can be seen in the motto for the modern Olympic Games: "faster, higher, stronger." Of course, Faster is first on the list. In 1896, in the marathon held at the Athens Olympics, a Greek runner, Spiridon Louis, won the race in 2:58:50. He might have finished sooner, but he stopped mid-race for a glass of wine.[15] While there are debates as to whether Louis ran the now-standard marathon distance of 42.195 kilometers (26 miles, 385 yards) or something shorter, the Olympic marathon has seen a steady increase in runners' speed. In 2008, Sammy Wanjiru, a Kenyan, set an Olympic record by running the prescribed distance in 2:06:32. Thus, in a span of slightly more than a century, the

Faster: Winning Times in Men's Olympic 100-meter Sprint, 1896–2012

Source: http://www.nytimes.com/interactive/2012/08/05/sports/olympics/the-100-meter-dash-one-race-every-medalist-ever.html?smid=fb-share.

winning male marathoner at the Olympics has cut his time by about a third.

The sprinters have also been getting Faster. At the Athens Olympics in 1896, American sprinter Tom Burke won the 100 meters in 12 seconds. At the London 2012 Olympics, Usain Bolt covered the same distance in 9.63. Put another way, Bolt ran the distance about 20 percent Faster than Burke did 116 years earlier.[16]

What accounts for Bolt's runaway success? There's no question that the Jamaican athlete was built for speed. His height (6 feet 5 inches, or 1.95 meters) and long legs give him an advantage over other sprinters.[17] Nor does Bolt lack confidence: "You can stop talking now, because I am a legend." But it's also apparent that all of his equipment, from his shoes to his singlet, are far Lighter, and far more precisely engineered, than the togs Tom Burke wore in Athens. Bolt also benefited from a running surface that was designed to make runners Faster. The track in London had an 8-millimeter-thick layer of diamond-shaped ridges beneath the top layer of the rubberized surface. As the *Wall Street Journal* explained, "By angling pieces of the subsurface, the track provided shock deflection both laterally and backward and forwards, propelling and stabilizing runners all at once."[18]

Faster, please: Driver Andy Green poses next to the Thrust SSC. This vehicle holds the world record for the fastest car. It was also the first to break the sound barrier. In 1997, it reached a top speed of 763 miles per hour (1,228 km/h) in Nevada's Black Rock Desert. The machine was powered by twin Rolls-Royce Spey 202 turbofan jet engines producing 50,000 pounds of thrust (roughly 110,000 horsepower).[19] The designers of this car are building a new vehicle, the Bloodhound SSC, which will use both jet engines and rockets. Their goal is to build a car that can travel at speeds in excess of 1,000 miles per hour.[20] *Source*: Thrustssc .com.

Innovation allows us to go Faster. And the more we push the boundaries of Faster, the more innovation we seek. While that push for innovation and speed often allows great human achievement, it can also result in grotesquerie. Few events offer a better example of human achievement, and human frailty, than the Tour de France.

7

FASTER LIGHTER DOPER

Modern athletes are in fact techno-human hybrids.
—Roger Pielke Jr.[1]

In his maniacal push for Faster, Tyler Hamilton lost so much weight and his skin got so thin, that his wife could see the outline of his internal organs.[2]

As an ambitious young cyclist eager to win the Tour de France, the world's most prestigious bike race, Hamilton starved himself for weeks at a time because he knew that the most effective way to go Faster was to get Lighter. Of course, Hamilton was also more than willing to cheat. At the Tour de France, innovation can be seen in the aerodynamic jerseys, bicycles, and training regimens of the riders. It can also, sadly, be seen in the extreme efforts that the cyclists have used to game the system, to use prohibited substances, and in doing so, gain an edge on their rivals.

Professional cycling has long been the world's dirtiest sport. No other athletic endeavor has such an ignominious history. In their never-ending effort to go Faster, cyclists like Hamilton, Lance Armstrong, Jan Ullrich, and dozens of others have been caught (or have admitted) using performance-enhancing drugs in order to gain an advantage over their competitors in the peloton. Doping was simply part of the program for those trying to win a spot on the podium. And to help them get there, the most sophisticated cycling teams focused on density.

Denser meant Lighter. Lighter meant Faster.

The metric for Faster Lighter Denser that Hamilton and other cyclists (and doping experts) have focused on is gravimetric power density, which is measured in watts per kilogram.

In his 2012 book, *The Secret Race: Inside the Hidden World of the Tour de France: Doping, Cover-ups, and Winning at All Costs,* Hamilton wrote about his rise to the top echelon of professional cycling during the 1990s and early 2000s, including his stint as Armstrong's teammate on the US Postal Service cycling team. The book explains that Armstrong's most-trusted adviser regarding when and how to cheat was an Italian physician named Michele Ferrari. Hamilton sought Ferrari's advice, too. The Italian doctor taught Hamilton that he should aim to spin his pedals Faster, or in cycling parlance, have a Faster cadence. But his most important advice was about density. Ferrari told Hamilton that "the best measure of ability was in watts per kilogram—the amount of power you produce, divided by your weight. He said that 6.7 watts per kilogram was the magic number, because that was what it took to win the Tour."[3]

Hamilton explained that Ferrari (who was paid more than $1 million by Armstrong between 1996 and 2006) "was obsessed about weight" because less weight for the same amount of watts resulted in a higher power-to-weight ratio.[4] And in cycling—just as it is with automobiles, airplanes, motorcycles, boats, and other machines—more power combined with less weight means more speed. Once Hamilton fully understood that to go Faster, he had to get Lighter, the starvation began. After long rides, he'd drink sparkling water to try to fool his stomach "into thinking it was full." He gorged on water and celery for days at time because "losing weight was the hardest but most efficient way to increase the crucial watts per kilogram number and thus do well in the Tour."[5]

Consider for a moment what that 6.7 watts per kilogram means. Some sports-medicine professionals believe that anything above 6.5 watts per kilogram is extraordinary and maybe not even possible for humans.[6] An amateur adult male cyclist in decent, but not top, physical condition can sustain an output of about 250 watts of power for an hour or so.[7] If you assume that an average male weighs 170 pounds (77 kilograms), that works out to about 3.2 watts per kilogram. My son Michael, now 18, is a competitive rower. He can produce about 300 watts over a 2-kilometer distance. He weighs about 150 pounds, or 68 kilos. Therefore, his power density is about 4.4 watts per kilogram. Thus, in theory, if Michael wanted to switch sports and aim his efforts at winning the Tour de France (an option I'm not encouraging), he would have to increase his power density by about 50 percent.

We can also understand the power output of elite cyclists another way. To win the Tour de France, a cyclist will need to have a gravimetric power density—recall Ferrari says it is about 6.7 watts per kilogram— that is about four times as much as a horse, which produces about 1.7 watts per kilo.

Faster Lighter at the Tour de France, 1903–2012

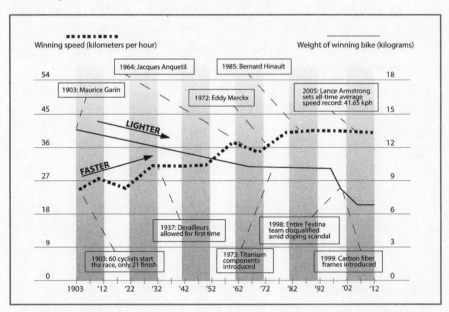

Between 1903 and 2012, the weight of the winning bicycle declined by 50 percent, and the average speed of the winning rider increased by 55 percent. The names in boxes at the top of the graphic are of notable winners and the year of their final win in the Tour. *Sources*: Tour de France, *Bicycle History: A Chronological Cycling History of People, Races and Technology*.

Hamilton's obsession with weight loss and power density is part of the century-long pursuit of Faster Lighter at the Tour de France. And that pursuit can easily be understood by looking at the weights of the bicycles used in the race.

In 1903, during the first Tour, just twenty-one riders finished the grueling race, which covered some 2,400 kilometers. The winner, Maurice Garin, had an average speed of nearly 25.7 kilometers per hour. Garin rode to victory on a steel-framed bicycle that weighed about 13.6 kilos (30 pounds).[8] The machine had a fixed gear, meaning that regardless of whether he was climbing or descending a steep hill, he could not stop moving his legs, nor could he gain any mechanical advantage by shifting gears. Instead, he had to simply grind his way through the entire race, all 2,428 kilometers (1,508 miles) of it.

Since the days of Garin, bicycles have steadily gotten Lighter. That light-weighting, along with other technologies, are allowing cyclists to convert more and more of their leg power into forward motion. In 1934, the Tour's winner, Antonin Magne, utilized a pair of lightweight aluminum-alloy rims, a technology that quickly became widespread.[9] Three years later, Tour officials allowed professional riders to use derailleurs, a now-ubiquitous technology that lets cyclists choose the gear ratios that are the best fit for the terrain.[10] Derailleurs allow riders to use different-size chain rings so that they can get maximum mechanical advantage regardless of whether the terrain is uphill, downhill, or flat.

By 1962, new metal alloys and better components (like brakes, shifters, wheels) allowed Jacques Anquetil to ride to victory on a bike that weighed 10.2 kilos (22.4 pounds). Ten years later, Eddy Merckx, perhaps the greatest cyclist in history, won the Tour on a bike that weighed 9.6 kilos (21.1 pounds).[11] In 1973, the Spanish rider Luis Ocaña became the first cyclist on the Tour to utilize components made from titanium, a metal that is as strong as steel but weighs about half as much.[12] Ocaña's bike weighed in at 8.5 kilograms (18.7 pounds).[13] The push to reduce the weight of bicycles flattened out for the next couple of decades. In 1993, the Spanish rider Miguel Indurain won the Tour on a bike that weighed 9 kilos (19.8 pounds).

A bicycle racer from the early 1900s. The basic design of the machine he's riding is much the same as modern bicycles. Note that his bike has no brakes. It also has a fixed gear, meaning the rider cannot stop pedaling while the machine is moving. *Source*: Library of Congress, LC-DIG-ggbain-04379.

In 2003, Lance Armstrong won the Tour on a carbon-fiber-framed bike that weighed just 6.6 kilos (14.5 pounds). That bike would turn out to be the benchmark as far as weight is concerned. In 2004, the International Cycling Union, apparently concerned about safety, decreed that bikes used in key competitions, including the Tour, could weigh no less than 6.8 kilos.[14]

Ever since the cycling union established a minimum weight for racing bikes, manufacturers and racers have been focusing on aerodynamic advantage. In 2004, the slippery shape of Armstrong's bike was estimated to reduce the needed power input by about 10 watts.[15] That's a significant savings for a Tour rider who must sustain outputs of 300 to 400 watts over a stage race covering more than 200 kilometers and lasting four hours or more. (During all-out sprints, the power output of elite cyclists can hit 1,000 watts or more.)[16]

In 2011, Cadel Evans, an Australian, became one of the oldest riders (34) to win the Tour de France. He rode to victory on a bike that utilized an aerodynamically tuned carbon-fiber frame equipped with electronic derailleurs.[17] Unlike their mechanical predecessors, which depended on the movement of a thin steel cable, the electronic version relies on a switch that activates a small solenoid, which then precisely moves the chain from one chain ring to another. That bike (which cost about $14,000) was also equipped with carbon-fiber wheels (eighteen spokes on the front wheel and twenty-four spokes on the rear) as well as a digital power meter that let Evans know exactly how many watts he was expending at any given time.[18] That same digital meter gave Evans a continuous readout of his speed, distance, heart rate, altitude, and pedaling cadence.

Compared to the primitive, heavy, fixed-gear machine that Garin used back in 1903, the bikes now used by the top racers in the Tour might as well be spaceships. The carbon-fiber frames, ultra-light wheels, and precision-machined components, combined with ever-more-aerodynamic shapes in helmets, clothing, and the bikes themselves, are allowing riders to go Faster. And Lighter Faster trumps heavier slower particularly in the Tour de France. But the human engines of the Tour—the cyclists like Hamilton, Armstrong, Evans, and the rest—impressive as they are, have never been able to match the output of the engines designed by people like Watt, Corliss, and Diesel.

Tour de Doper

The world's most famous cycling event—indeed, the entire history of bicycle racing—has been plagued by doping. Of course, many other sports, from sprinting to baseball, have seen their share of cheaters who rely on banned substances. But when it comes to pharmacological chicanery, cycling has no peer.

In the push to go Faster, cyclists have used strychnine, chloroform, and cocaine during races. Others have used ether or alcohol. In the modern era of the Tour, cyclists have used anabolic steroids, EPO, testosterone, human growth hormone, female hormones, insulin, and blood transfusions. Herewith a few examples of the doping at the Tour:

1949: A rider named Fauso Coppi admits using amphetamines during the Tour.

1962: Nearly two dozen riders become ill during the Tour, likely because of morphine doping.[19]

1965: During a TV interview, Jacques Anquetil, who won the Tour four years in a row, dismissed concerns about drug use, saying simply, "Leave me in peace. Everybody takes dope."[20]

1987: Two riders, Guido Bontempi and Dietrich Thurau, failed drug tests after early stages in the race and are given time penalties of 10 minutes apiece.[21]

1997: The Tour is won by Jan Ullrich, a German, who goes on to win both gold and silver medals in cycling at the Olympic Games in 2000. In 2013, Ullrich admitted that he had cheated, saying, "Almost everybody back then took performance-enhancing substances."[22]

1998: The Tour becomes known as the "Tour de Farce" after a member of the Festina team is caught with anabolic steroids, growth hormones, and a variety of masking agents. A few days later, the entire Festina team is expelled from the Tour.[23]

1999–2005: Lance Armstrong wins the Tour seven consecutive times.

2004: Lance Armstrong tells Oprah Winfrey, "I have to win at all costs."[24]

2006: American rider Floyd Landis wins the Tour but tests positive for doping and is stripped of his title. In 2010, Landis admits that he had cheated and says he began using performance-enhancing drugs in 2002, when he was a member of the US Postal Service team. The star of that team was Armstrong.[25]

August 2012: Jonathan Vaughters, a prominent American cyclist, admits that he had doped during his career. In an op-ed published by the *New York Times*, Vaughters wrote that doping can provide "that last 2 percent" that can "keep your dream alive, at least in the eyes of those who couldn't see your heart." He continued, pointing out that an extra "2 percent of time or power or strength is an eternity." In the Tour, he said, "2 percent is the difference between first and 100th place in overall time."[26]

August 24, 2012: The US Anti-Doping Agency issues a lifetime ban from competition on Lance Armstrong in all sports that follow the World Anti-Doping Agency code. Two months later, the International Cycling Union strips Armstrong of all seven of his Tour titles.[27]

January 2013: Armstrong admits that he'd taken banned substances, and/or had done blood doping, in all of his wins at the Tour de France.[28]

8

THE ENGINES OF THE ECONOMY

For all of the talk about what creates economic growth and prosperity, there is one unassailable truth: the engines of the economy are engines.

For centuries, businessmen, engineers, inventors, and everyday working people have been finding ways to harness more energy so that we can do more work Faster Cheaper. Our ability to do work, and therefore, create economic growth, largely depends on our ability to convert energy of whatever type—heat, kinetic, chemical—into motion. By doing so, we are able to produce the power we need to perform the work at hand. Engines convert energy into motion that can be used for doing work. As my friend Stan Jakuba has put it, the more we increase *the energy flow* through those engines, the more power we get and the more work we can do.* And the more work we can do of whatever type, the more work we want to do.

The entirety of the Industrial Revolution, through the Age of Steam, the Age of the Automobile, and today's Information Age, can be described as the push to derive more useful power from the energy we consume. We are, all of us, power hungry. We want Cheaper watts. Always. Everywhere. And the push to produce Cheaper watts has focused on our ability to more efficiently convert the joules of energy we put into our engines into more watts. Proving that point requires us to look back a few centuries to see how engine technology has developed, and more particularly, how we have made our engines Denser—how we have continually sought engines with higher power densities.

---

* Energy, measured in joules, is the ability to do work. Power, measured in watts, is the rate at which work is done.

A waterwheel on the Orontes River in Syria. It provided water
for agricultural and urban use. A wheel of this size likely produced
the equivalent of about 40 horsepower. *Source*: Library of Congress,
LC-DIG-matpc-06756.

Of all the artifacts of antiquity that demonstrate the human desire
for power, few are bigger, or more beautiful, than the waterwheels
of Hama, a region in west-central Syria. Built in the Byzantine era,
the waterwheels are marvels of engineering. Fashioned primarily from
wood and stone, they lifted water from the Orontes River into aqueducts
that carried the water to farms, mosques, and urban centers. About
seventeen of the Syrian waterwheels are still in existence, a fact that tes-
tifies to the durability of their construction. Standing as much as 66 feet
(20 meters) high, waterwheels like those in Hama were capable of moving about 1,270 gallons (4,800 liters) of water per minute.[1] While there
are no reliable estimates of how much the waterwheels weighed, we can
estimate their horsepower by comparing their water-moving capabilities
with that of modern pumps. Doing so indicates that waterwheels like

the ones at Hama likely produced roughly 40 horsepower (about 29,000 watts), a modest sum by today's standards, but a remarkable feat for the thirteenth century.[2]

The waterwheels at Hama are only one example of a technology that dates back to the ancient Greeks. But before we delve further into waterwheels, and we will, let's look at the only other power options that were available to humans before the Age of Steam: human muscle, animal muscle, or wind energy.

Sailing vessels were used to move goods on rivers and along coastlines, and windmills were useful in locations that had reliable wind. But when it came to doing the everyday work of civilization—transporting goods, digging in the fields, processing grain, or moving water—we had to do it ourselves or put a harness on draft animals. And for millennia, that's what we did.

Humans have powerful brains but relatively weak bodies. A person in average physical condition can produce somewhere between 60 and 120 watts of power while doing moderate to strenuous work. (Elite athletes, of course, can produce far more than that.)[3] Rather than do all the work themselves, humans used their big brains to innovate: They made harnesses that allowed them to use oxen, water buffaloes, or cows to pull carts and plows as well as to mill grain. Harnessing cattle allowed humans to cultivate land at least three times Faster than what could be done by a peasant armed with a hoe.[4] And while cattle were useful, they were not as powerful as a horse and could produce only 300 to 400 watts (about one-half horsepower) of power. Further, their slower gait meant they could only travel about two-thirds as fast as a horse.

Draft horses were far superior to cattle in power output, speed, and stamina. For short durations, horses could pull as much as 35 percent of their body weight, which in some cases meant they could produce about three horsepower (about 2,200 watts). Horses were durable, relatively easy to manage, and with the advent of the collar harness, which became prevalent in Europe by about the ninth century, they could pull heavy loads comfortably for many hours. But heavy draft horses also had to eat. Horses can survive on grass-only diets. If they are worked hard on a daily basis in a harness or under a saddle, they need better diets. That means grain. The results were obvious: as more horses were put to work, their need for grain increased. And that put them in direct competition with humans. By the early 1900s, as much as 20 percent of all US farmland was being used to cultivate grain solely for horse feed.[5]

For centuries, horses were the engines of the economy, but they were never very powerful. To illustrate, let's return to the metric of power density. Calculating the power density of a horse is fairly straightforward. Assume an average-size horse weighs 1,000 pounds (call it 450 kilos). A horse can produce 746 watts (one horsepower). Given those numbers, the math is easy: 746 watts divided by 450 kilograms gives us a power density of about 1.7 watts per kilogram.

The limitations of human muscle, cattle, and horses led inventors and engineers to design waterwheels and windmills in order to turn the available kinetic energy of the rivers and the wind into useful power. The Romans used waterwheels most commonly for small-scale milling of grains.[6] The earliest use of windmills likely occurred in about the tenth century A.D., in Seistan, a region of eastern Iran. The windmills were used to pump water for agriculture.[7] Around 1300, windmills began to proliferate in medieval Europe and were used for moving water, milling grain, and other purposes.[8] Over time, windmills grew to be quite large, as the builders began to understand that adding height to the windmill was the best way to increase its power. That trend continues today. Many of the latest turbines stand about 500 feet (150 meters) high.[9] But even as the European windmills proliferated through the 1700s, they were still relatively modest in terms of power output, with even the most efficient of the machines able to produce perhaps 13 horsepower, or about 10,000 watts.[10]

While windmills were popular, particularly in northern Europe, the advent of the Industrial Revolution was made possible by waterwheels, which were the prime movers of choice for factories that churned out everything from cloth to firearms. But while the waterwheels were effective at converting the kinetic energy of flowing rivers and streams into rotating mechanical power, they were always at risk from both drought and flood. Too little water, or too much, imperiled the factories

This photo of an Amish farmer working his fields near Lancaster, Pennsylvania, taken in about 1980, shows a method of working the land that has lasted for centuries. This team of horses can generate 6 to 8 horsepower for several hours. For short bursts, it can probably produce two to three times that amount. *Source*: Library of Congress, LC-DIG-highsm-16027.

and the livelihoods of the workers in them. Nevertheless, inventors and entrepreneurs teamed up to create entire industrial ecosystems that were powered by waterwheels. Author Charles R. Morris in his excellent 2012 book on the American Industrial Revolution, *The Dawn of Innovation*, points to the development of the Locks & Canal Company in Lowell, Massachusetts, during the early 1800s as a pivotal period in American history. The company diverted the Merrimack River in order to create what Morris calls a "hydraulic power utility." By the mid-1830s, the utility was providing waterpower to twenty-five textile mills, as well as a variety of other operations. By the late 1840s, the Locks & Canal Company was operating a network of canals 17 miles long. The arrangement of factories in the Lowell area was, writes Morris, "by far the greatest industrial development in the country, and its impact on machining, metalworking, and other industrial technologies is hard to overestimate."[11]

Although the waterwheels fed the factories that sparked the Industrial Revolution in America, they were never very efficient. Morris points to one particular waterwheel designed by Henry Burden, an American inventor. In 1851, Burden built what was then the world's most powerful waterwheel in Utica, New York. The machine was 62 feet in diameter and weighed 250 tons. While it proved to be extremely durable, operating for some fifty years, its maximum power output was about 300 horsepower.[12] That's a remarkable amount of power for the 1850s. But when looked at in terms of gravimetric power density, it only produced about 1 watt per kilogram.

The desire for more power, along with the limitations of human muscle, draft animals, wind energy, and waterwheels led inevitably to the development of the steam engine. The first steam engine used in an industrial setting was built by Thomas Newcomen (b. 1664, d. 1729), an Englishman. First deployed in 1712, it was used to pump water out of a tin mine. While Newcomen's engine was ingenious, it burned copious quantities of coal. Furthermore, the engine was built to very low tolerances. The boiler used in the Newcomen engine was limited to pressures of about 2 pounds per square inch (13.8 kilopascals).[13] That's not very much when you consider that compression levels inside the cylinders in modern automobile engines max out at about 1,000 pounds per square inch (6,895 kilopascals).[14]

Pivotal improvements to Newcomen's design were made by a Scotsman, James Watt, whose last name has become synonymous with power. Watt coined the term "horsepower." Today, Watt's name lives on as the standard unit of power, the watt. Watt (b. 1736, d. 1819) assessed the Newcomen engine and estimated that about 75 percent of the fuel it used was being wasted in the reheating of the engine's cylinder after the cylinder had been cooled to create a vacuum. Watt saw that a separate condenser unit could be employed so that the cylinder, and the piston inside it, stayed hot, and thereby saved energy. Watt saw a

James Watt's improvements to the steam engine helped ignite the Industrial Revolution. The SI unit for power, the watt, is named for him. *Source*: Wikimedia Commons.[15]

business opportunity in a simple idea: make engines that were "cheap as well as good."[16]

In 1774, Watt, who had been making his living as a surveyor, teamed up with Matthew Boulton, an industrialist who had a knack for both business and promotion.[17] Over the next two decades, their firm, Boulton & Watt, sold hundreds of engines.[18] And while their engines were reliable, safe, and relatively efficient, they were not overly powerful. A typical Boulton & Watt engine from the early nineteenth century was capable of producing about 24 horsepower (17,900 watts). But it weighed about 2 tons (1,818 kg) giving it a gravimetric power density of 9.8 watts per kilogram. That was an enormous improvement when compared to the power density of many waterwheels. It was also nearly six times the gravimetric power density of the average horse. But unlike a horse, the steam engine could be worked around the clock, produced no manure, and didn't require grain, meaning it didn't compete with humans for available farmland.

Boulton & Watt engines unleashed the Age of Steam and led to a surge in productivity and commerce.[19] Unlike waterwheels, which had to be placed close to rivers, steam engines could be put almost anywhere. Furthermore, they could be used for more than just manufacturing; they could be used for transportation. That use proved to be critical because without transportation, there is no commerce.

Replacing sails with steam meant humans could travel farther Faster Cheaper than ever before. In August 1807, after years of prototypes and failures, inventor Robert Fulton (born in Little Britain Township, Pennsylvania, in 1765) launched the *Clermont*, the first successful commercial steam-powered boat.[20] Fulton's boat (powered by a Boulton & Watt engine) traveled the Hudson River, carrying passengers between New York and Albany.[21] At the time that Fulton launched the *Clermont*, the sail-powered sloops carrying passengers and freight between the two cities were taking about a week.[22] Fulton's boat made the same voyage in about thirty-six hours.[23]

A few years after launching his service in New York, Fulton, who had teamed with one of New York's richest men, Robert Livingston, began operating steamboats in the western United States. In 1811, Fulton's boats began carrying passengers from Pittsburgh to New Orleans, a journey of about 2,000 miles, via the Ohio and Mississippi Rivers.[24] Within a few years, steamboats proliferated on the Mississippi and other rivers, and those boats played a critical role in the opening of the American West.

Replacing sails with steam meant Faster travel on the water, but it was the use of steam in factories and on rails that supercharged the Industrial Age. The more steam engines that were produced, the more factories there were that relied upon them. As engine production capacity grew, so too did advancements in metallurgy, lubrication, machine tools, punches, presses, hammers, and all the other technologies needed to make engines that were Smaller Faster Lighter Denser Cheaper than their predecessors. More engines begat better engines, and the rapidly spreading steam-powered railroads and ships fed the symbiosis.

The first railroad that carried people and freight on a regular schedule began service in 1825, carrying passengers and coal from the British coal town of Darlington to the port at Stockton. As Jeff Goodell writes in his book *Big Coal,* the railroads were a key invention that led to more coal production because, "in effect, coal hauled itself." Goodell points out the mutually beneficial relationship between the shippers and the fuel producers: "The partnership of railroads and coal created a kind of perpetual motion machine: better transportation meant cheaper, wider distribution of coal, which fed the growth of steel mills and steam power, which in turn further increased the demand for coal."[25]

It wasn't just steel and coal. According to William Rosen, the Industrial Revolution was also fueled by another commodity: cotton. Coal-fueled steam engines drained the mines that produced yet more coal. Coal fueled the forges and furnaces that produced the steel needed to produce the engines that were put into the steamships that carried raw cotton to the British Isles, where it was spun into cloth by steam-powered mills. The finished cloth was then carried to market on steam-powered railroads and ships. The steam engine, writes Rosen in his 2010 book, *The Most Powerful Idea in the World,* created a "perpetual innovation machine in which each new invention sparked the creation of a new one, ad—so far, anyway—infinitum."[26]

While the three spokes of steel, coal, and cotton drove the early stages of the Industrial Revolution—and coal's dominance lasted more than a century—that fuel's dominance of the transportation sector would eventually yield to an even better energy source: oil.[27]

Just as the early days of the coal industry were tied to the railroads, so too were the early days of the oil industry. With no interstate pipelines available to carry crude oil or refined products, John D. Rockefeller, the founder of Standard Oil, understood that Faster Cheaper rail service was essential if he was to make oil Cheaper for his customers. In the late 1860s, Rockefeller began investing in railroad tanker cars, a move that saved him the cost of building barrels to hold his products.[28] The oil baron's control over the Cleveland-area refining market allowed him to negotiate favorable shipping rates with the railroads. As Ron Chernow notes in his biography of Rockefeller, *Titan: The Life of John D. Rockefeller, Sr.,* the railroads were ready and willing to help Rockefeller because he was moving so much freight. By consolidating their operations around Rockefeller's oil shipments in the late 1860s, the railroads could move freight Faster and "reduce the average round-trip time of their trains [from Cleveland] to New York from 30 days to 10 and operate a fleet of 600 cars instead of 1,800." With that deal, writes Chernow, "the railroads acquired a vested interest in the creation of a gigantic oil monopoly that would lower their costs, boost their profits, and generally simplify their lives."[29] (Railroads continue to be a major transporter of oil. In 2012, about 4 percent of all North American crude oil production—about 400,000 barrels per day—was being moved by rail.[30] Russian oil producers are also heavily dependent on railroads for transportation.[31])

Of course, the railroads have done far more than haul coal and petroleum. They also played a key role in the development of the American West. Throughout the 1800s and early 1900s, entrepreneurs and inventors built bigger and bigger steam engines. The adjacent picture shows a locomotive from 1865. It weighed about 30 tons (27,000 kilograms) and produced about 500 horsepower (373,000 watts), giving it a gravimetric power density of about 14.2 watts per kilogram.*

The Age of Steam—in factories, on rails, and on boats—was one of continuing innovation. And while the Boulton & Watt engine was reliable and durable, it used a low-pressure boiler. That meant that it weighed more and produced less power than comparable engines that used high-pressure designs. Among the early adopters of the high-pressure steam engine were a pair of savvy engineers, Daniel French and Henry M. Shreve, who began operating riverboats on the Mississippi River in 1815. Although the high-pressure steam engines were more dangerous—their boilers had an unfortunate tendency of exploding—the high-pressure engines had higher gravimetric power density than the Boulton & Watt design.

---

* Note that the density figure for the engine itself is a little low as it counts the weight of the entire locomotive—the wheels, etc.—while the other prime movers discussed here generally count only the weight of the engines themselves.

This locomotive, which was used to carry Abraham Lincoln's body after his assassination in 1865, likely produced about 500 horsepower (373,000 watts). For comparison, the 2013 Ferrari F12 produces 730 horsepower (about 544,000 watts). *Source*: Library of Congress, LC-DIG-ppmsca-23855.

In 1816, French and Shreve began operating a boat powered by a high-pressure steam engine of their own design. That engine produced 100 horsepower and weighed 5 tons (4,545 kg).[32] Simple math shows that the French and Shreve engine had a gravimetric power density of 16.4 watts per kilogram. That was a significant increase over comparable Boulton & Watt engines of the time, which, as mentioned above, had power densities of about 9.8 watts per kilogram. That big boost in power-to-weight ratio was of critical importance to steamboat pilots, who often needed extra bursts of power when navigating treacherous waters.

The Age of Steam likely reached its acme in Philadelphia in 1876, when an American, George Corliss, showcased one of the most famous steam engines of all time. Corliss (b. 1817, d. 1888) was a native of up-state New York, and like Watt, he had a genius for understanding and improving the mechanics of the steam engine.[33] He designed valves and a valve-management system that improved the steam engine's fuel efficiency by about 30 percent. After winning the gold medal at the Paris Exposition of 1867, Corliss became renowned as one of the world's foremost builders of steam engines. Charles Morris explains that Corliss's engines were so smooth and dependable that they became "almost a must-have" in the textile industry, which needed engines that could run their looms at constant speeds.[34]

On May 10, 1876, 190,000 visitors flooded into the fairgrounds for the Centennial Exhibition in Philadelphia, which celebrated the hundredth anniversary of the Declaration of Independence.[35] On that day, President U. S. Grant and Brazil's Emperor Dom Pedro started the giant 1,400-horsepower (1 megawatt) Corliss steam engine that dominated the center of Machinery Hall. The massive machine stood 45 feet high and was the prime mover for some eight hundred machines arrayed inside the hall. The Corliss engine powered the machines through shafts that totaled more than a mile in length. The machine's size and power left visitors slack-jawed. Journalist William Dean Howells, editor of the *Atlantic Monthly*, wrote that the machine was "an athlete of steel and iron with not a superfluous ounce of metal on it." Howells continued, "the mighty walking beams plunge their pistons downward, the enormous flywheel revolves with a hoarded power that makes all tremble, the hundred life-like details do their office with unerring intelligence."[36]

Although the total weight of the Corliss engine is not available, we can approximate the machine's power density by using the weight of its flywheel: 56 tons, or 50,800 kilograms.[37] With an output of 1,400 horsepower or just over 1 million watts, the gravimetric power density of the Corliss engine was about 20 watts per kilogram, or about twice the density of the Boulton & Watt engines that had dominated the industrial sector seven decades earlier.

While Corliss's engine was a marvel of its day, the steam engine's reign would not, could not, last. The push for Denser, more powerful engines led to the development of the internal combustion engine. In 1886, a German inventor, Karl Friedrich Benz, registered a patent on a three-wheeled automobile powered by a gasoline-fueled internal combustion engine.[38] Benz's vehicle used a 1-cylinder engine that weighed 211 pounds (96 kg) and produced 500 watts of power.[39]

An illustration showing the giant Corliss steam engine
at the 1876 Centennial Exposition in Philadelphia. The
machine generated about 1,400 horsepower, or roughly
1 megawatt. *Source*: Library of Congress: LC-USZ62–96109.

The Age of Steam was yielding to the Age of the Automobile. While Benz kick-started the era, an American, Henry Ford, brought motive power to the masses. In doing so, Ford offered consumers a durable internal-combustion engine that was Smaller Lighter Denser than the steam engines that birthed the Industrial Revolution.

While both engines depend on the burning of hydrocarbons, the steam engine's process is indirect: fuel is burned in a boiler, and the heat is used to create steam, which is then routed to a cylinder, where it drives a piston to create motion. By contrast, internal combustion engines take direct advantage of the explosive power of hydrocarbons inside the cylinder. By doing so, internal combustion engines are able to produce more power from a smaller package, as can be seen in the gravimetric power density that was available in Ford's Model T, which debuted in 1908 at a cost of $850.

The original Model T was equipped with a 2.9 liter engine that produced 22 horsepower (about 16,400 watts) and weighed about 300 pounds (136 kg). The result: gravimetric power density of nearly 121 watts per kilogram. That power density was 73 times that of a horse, 12 times that of the Boulton & Watt design and about six times that of the engine Corliss had introduced in Philadelphia three decades earlier.

The engine in the Model T launched a mobility revolution. In 1908, Ford built fewer than 11,000 of the cars. By 1916, production was 600,000.[40] By 1927, Ford was making a new Model T every 24 seconds. That same year, production of the Model T was finally halted. In all, some 15 million copies of that vehicle had been sold to the public.[41]

While the Model T marks a pivotal moment in the history of the Age of the Automobile, the quest for higher power density in internal combustion engines was only beginning. Ford's assembly line idea, combined with the advent of electricity, and more advanced manufacturing techniques, allowed engineers to design and build Smaller Faster Lighter Denser Cheaper engines that could be used in everything from automobiles and airplanes to lawn mowers and Weedwhackers.

The development of the internal combustion engine launched a race for horsepower that's ongoing to the present moment. Mobility was, and is, the name of the game. The quest for Faster has been propelled

Denser: Measuring Power Density from Horses to Jet Engines

| = Gravimetric power density of an average horse |

| World-class cyclist (2000s) | Boulton & Watt steam engine (early 1800s) | Corliss Centennial Exhibition steam engine (1876) | Model T engine (1908) | Ford 1-liter EcoBoost engine (2012) | GE GEnx-1B jet engine (2013) |

This graphic compares the gravimetric power density of an average horse, which is 1.7 watts per kilogram, with other power sources. Automobile shown is a Model T. For source data, please see Appendix C.

by engines fueled by the growing abundance of cheap refined oil products. (And as mentioned earlier, those Cheaper motor fuels were being made available thanks to Faster Cheaper drilling technologies like the roller-cone bit.) By 1917, with World War I raging, American aviators were flying the Curtiss JN-4, known as the Jenny, which was equipped with a 90-horsepower (about 67,000 watts) engine, while soldiers on the ground were riding in Mark VIII tanks powered by 300-horsepower (about 224,000 watts) engines. Less than three decades later, American flyers were dominating the World War II skies over Europe thanks to the P-51 Mustang, a high-performance fighter that was equipped with an engine that produced 1,695 horsepower (1.26 megawatts).

In less than 70 years, American industry had gone from producing the Corliss engine, which weighed more than 50 tons and produced 1,400 horsepower, to producing an airplane that produced 21 percent more power and yet weighed just 6 tons. Furthermore, that aircraft could travel at speeds of more than 400 miles per hour.[42] That was many times Faster than any of the steam-powered locomotives of the late-nineteenth and early-twentieth centuries, which usually had top speeds of about 60 miles per hour.[43]

The quest for Faster transportation meant designers had to create more powerful engines. Thanks to better alloys, more advanced machine tools, better lubricants, and closer tolerances, they have done exactly that. Modern engines use variable valve timing, fuel injection, and onboard computers to wring more power out of each joule of energy consumed.

In the span of two centuries, modern society has gone from depending on Boulton & Watt steam engines capable of producing about 10 watts per kilogram to using jet turbines that are 1,500 times as powerful per unit of weight. A century ago, consumers were astounded that Ford could produce Model T engines that produced 22 horsepower. And at 121 watts per kilogram, those Model T engines sparked a world-changing advance in mobility. Today, consumers can easily purchase automobiles powered by engines that produce more than 900 watts per kilogram. If they want to see cars that can go really fast, they can watch Formula One cars powered by engines with power densities of about 5,900 watts per kilogram.[44]

The push to wring more and more power out of Smaller Lighter engines has been ongoing since the days of Newcomen and Watt. And it will continue. In 2012, according to Dennis Huibregtse of Power Systems Research, some 222 million engines were manufactured.[45] That works out to about one new engine per year for every thirty-one inhabitants on the planet. We rely on those devices to power everything from hedge trimmers to supertankers.

The engines of the economy are engines, and every year those machines are getting Smaller Faster Lighter Denser Cheaper.

SMALLER FASTER INC.

FORD

Official Name Ford Motor Company

Web site http://www.ford.com

Ownership Publicly traded, NYSE: F

Headquarters Dearborn, Michigan

Finances Market capitalization: $49.3 billion[46]

2012 Revenue $136 billion[47]

From his very first vehicles, Henry Ford was obsessed with Faster. In May 1902, Ford built the 999 Racer. It had a wooden chassis and massive engine that displaced 1,155 cubic inches (18.9 liters) and produced about 70 horsepower. In 1904, the car set a world speed record: 91.4 miles per hour.

Few companies have lasted as long as Ford Motor Company has. Fewer still have been as successful. Ford's enduring success can be attributed to its continued pursuit of Smaller Faster Lighter Denser Cheaper.

Let's consider Cheaper. In 1908, when Ford launched the Model T, consumers had only one color choice: black. And the price, while attainable for many people, was still fairly steep: $850. In current-day dollars, that vehicle would cost more than $20,000.[48] By comparison, the lowest-priced Ford being sold in the United States is the Fiesta. A search done in mid-2013 on Cars.com found a brand-new, four-door, 2013 Ford Fiesta SE (in "violet gray") with a five-speed manual transmission, cruise control, front disc brakes, power windows, a CD player, and a raft of airbags—first

1902: Henry Ford (right) stands next to the 999 Racer and driver Barney Oldfield. The engine on the vehicle produced about 70 horsepower (52,220 watts) from 18.9 liters. A modern Ford engine, the 1-liter EcoBoost, has far greater power density. It produces about 75 percent more power from an engine that displaces about 5 percent of the volume required by the 999. *Source*: Wikimedia Commons.[49]

and second row curtain head, passenger-side, and driver-knee— could be purchased for a shade less than $12,000.[50] Put another way, over the course of a century, the base-model Ford automobile has declined in price by about 40 percent compared to the first of the Model Ts. (Ford quickly maximized production, which allowed the company to cut prices. By 1915, a Tin Lizzy could be had for $440.)[51]

The price decline from 1908 to today would be far greater if Ford Motor were still building a vehicle as simple as the Model T, which, by the way, had a top speed of about 45 miles per hour (72 km/h). The modern Fiesta costs less than the original Model T, and it provides safety, comfort, and speed advantages that would have left automobile buyers of a century ago agog. Compared to the Model T, the Fiesta can travel about twice as fast, with far greater reliability, and it even has a cup holder for your half-caf mocha latte while you cruise to the latest sounds from Tallest Man on Earth or maybe some vintage J. J. Cale.

That Fiesta is Smaller than the Model T. Its engine is Smaller (1.6 liters versus 2.9).[52] And the wheelbase is about two inches shorter (98 inches versus 100 inches).[53]

Denser? You bet. Over the past century, Ford has dramatically increased the power density of its engines. In 2011, the company unveiled a 3-cylinder turbocharged 1-liter engine, the EcoBoost, which can produce 123 horsepower (91,758 watts).[54] That's about 16 times as much power per liter of displacement as the engine in the first Model T, and more than 30 times that of the 999 Racer. Even more remarkable is this: the 1-liter EcoBoost is more than two times as fuel-efficient as the engine used in the original Model T.[55]

Lighter? Absolutely. The 1-liter EcoBoost engine is about 28 percent (39 kg) Lighter than the one that was used in the Model T. The result is a big boost in gravimetric power density: the EcoBoost produces about 946 watts per kilogram, nearly eight times as much as what was produced by the Model T. The push for Lighter has extended into nearly every product that Ford makes. In 2011, a high-ranking manager from a rival carmaker told me that Ford was the "most innovative company in lightweight manufacturing." In early 2014, Ford said it would begin making the body of its F-150 pickup from aluminum, a move that will make the vehicle about 700 pounds Lighter.

Faster? Oh my, yes. In 2012, the EcoBoost engine was named the International Engine of the Year, a distinction that came, in part, because one of its key components rotates at astounding speed.[56] Turbochargers have been used to boost power output in engines for decades. They do so by forcing more air into the cylinder prior to combustion. The impeller in a turbo commonly operates at speeds of 60,000 to 100,000 revolutions per minute.[57] The impeller in the EcoBoost tops out at 248,000 rpm.

Smaller Faster Denser: Volumetric Power Density in Ford Engines, 1902–2011

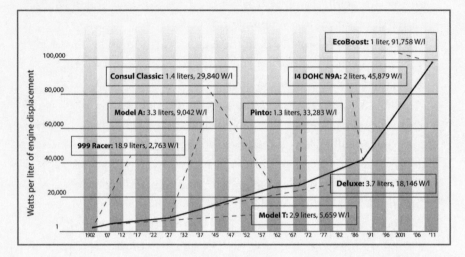

Over the course of a century, the engineers at Ford Motor Company have dramatically increased the volumetric power density of the company's engines. In 1902, the 999 Racer was able to produce less than 3,000 watts per liter of engine displacement. By 2011, with the introduction of the EcoBoost, Ford's engines were producing more than 90,000 watts per liter, a thirtyfold increase.

Faster components allow more power to be produced from a Smaller space. A fundamental rule in physics is that power is equal to torque times speed. Torque is related to mass (weight), and given that Ford's engineers wanted to reduce the engine's weight, they couldn't add heavy components. That left them with the challenge of engineering a turbocharger that could spin Faster than previous versions while keeping it light. The solution: Ford's engineers crammed the turbo into a unit that's about the size of an orange. In addition to the turbo, Ford's designers added other go-fast technologies like direct injection and variable timing for the inlet and exhaust camshafts.[58]

The turbo is Smaller. So is the engine itself. The outline of the 1-liter EcoBoost's crankcase nearly fits on a sheet of office paper, allowing Ford to build cars that go Faster. In 2012, on a French track, the company set sixteen world speed records with a handful of Ford production cars equipped with the 1-liter EcoBoost. The records were set for vehicles in their engine class and included highest average speed over 10 kilometers and highest average speed over 24 hours: 106.2 mph (171 km/h).[59]

The EcoBoost not only proves the durability of innovation and the push for density, it also shows that gasoline-fueled engines are going to be around for a while. And if Ford keeps innovating, the company might even be around for another century.

9

FROM ENIAC
TO iCLOUD

SMALLER FASTER COMPUTING

For Apple devotees and the herd of reporters inside the Moscone Center in San Francisco, the June 6, 2011, presentation from Steve Jobs at the Worldwide Developers Conference was familiar: the Apple CEO and design visionary was presenting yet another round of products and services that would further cement the company's position as one of the world's most innovative technology providers.[1]

But the 2011 meeting in San Francisco soon became notable for another reason: it would be Jobs's next-to-last public appearance as the CEO of the world's most famous technology company.[2] The following day, he appeared before the Cupertino City Council to pitch the council on the design for the company's new campus.[3] Two months later, the tech titan resigned as Apple's CEO, and two months after that, he was dead, felled by pancreatic cancer.

At the time of the meeting, the biggest news from the Worldwide Developers Conference wasn't Jobs's health (which was constantly being scrutinized); it was that the visionary entrepreneur was introducing iCloud, a service that would allow users to "automatically and wirelessly store your content in iCloud and automatically and wirelessly push it to all your devices." Jobs explained that Apple was "going to move the digital hub, the center of your digital life, into the cloud."

Other tech companies had been promoting their plans for the "cloud"—the industry's name for the network of data centers that have become the digital brains of our society—but Apple's move was different. Other companies had launched phones and tablet computers. Apple produced category killers like the iPhone and the iPad, and by announcing its move into the cloud, Apple was providing an endorsement of the concept. Apple was making ubiquitous computing real. It was removing the task of information storage out of the hands of consumers— or rather it was moving data off of the flash drives and hard drives inside mobile phones, iPads, and laptops—and moving it into massive data centers crammed with servers. Making that happen was no small chore. Apple's iCloud data center, in Maiden, North Carolina, Jobs explained, was going to be one of the biggest in the world, with some 500,000 square feet (46,451 square meters), making it about five times as large as an average Walmart discount store.[4]

Apple was aiming to dominate cloud computing for consumers despite lots of competition. Amazon, the giant online retailer, was selling its service, Cloud Drive. Google was marketing Google Drive, and Microsoft was offering SkyDrive.[5] Near the time of Jobs's announcement, each of those companies was offering about five gigabytes of storage in the cloud for free. And in mid-2013, Flickr, the photo-sharing site, began offering users 1 terabyte of storage space, which was enough to store more than 500,000 photos—all for free.[6]

You can't get any Cheaper than free. Just consider how much Cheaper computing has gotten over the past few years. In 1997, a consumer who needed 5 gigabytes of hard drive storage would have had to pay about $450.[7] That money purchased only the hard drive itself with no computing capability around it. Today, that volume of storage is free. Better yet, for consumers, all of that data storage comes with no power cords, cables, or other hardware that can be tripped over, spilled on, or otherwise damaged. Our ability to keep the latest song by Lady Gaga as well as that cute video of Grandma Val's cat is not only free; the storage itself has become invisible and, for the person using it, weightless.

No other sector better demonstrates the march of Smaller Faster Lighter Denser Cheaper than computing. In 2013, the inventor and author Ray Kurzweil told the *Wall Street Journal*, that "a kid in Africa with a smartphone is walking around with a trillion dollars of computation circa 1970."[8]

Faster Cheaper: The Volume of Digital Data Created and Shared, projected to 2015

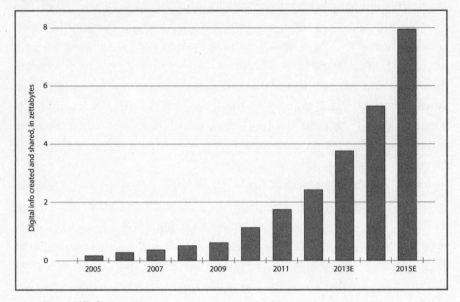

Source: IDC, 2011.

Cheaper electronics—whether they are desktop computers, land-line phones, or smart phones—are allowing humans to exchange gargantuan quantities of information. Thanks to Faster networks, the volume of that information is growing at a staggering pace. Between 2006 and 2011, the volume of digital information—from YouTube videos to tweets—that was created and shared grew ninefold, to some 2 zettabytes of data.[9] For reference, a zettabyte is 1 sextillion bytes, or 1 trillion gigabytes, or 10^{21} bytes. Comprehending that volume of information requires a bit of work. To put it in perspective, consider this: if we could store the Library of Congress's entire collection in digital form, it would fill roughly 200 terabytes of hard-drive space.[10] (A terabyte is 10^{12} bytes.)

Recall that in 2011 we humans exchanged about 2 zettabytes of data. And by 2015, forecasters from the consulting firm IDC expect we will be exchanging 8 zettabytes. If that happens, by 2015 we will be exchanging annually about 40 million times as much data as exists in the Library of Congress.[11] And much of that data exchange will be happening on a global fiber-optic network that is exchanging photons traveling at 300 million meters per second.

The world is now networked, and it's getting more wired every day. In 2013, Google announced that it was planning to build wireless networks using high-altitude balloons and airships so that it can provide connectivity to people living in rural areas of Africa and Asia.[12] This surge in both data exchange and connectivity is a direct result of the push toward Smaller Faster. In 1965, Gordon Moore, the cofounder of Intel, famously declared that computing power—measured by the number of transistors placed on an integrated circuit—would double every two years.[13] That declaration, known as Moore's Law, has been proven right so far. Intel's latest process can print individual lines on chips that are a thousand times thinner than a human hair.[14] Pursuing such density on microprocessors means more computing power. And computing power is like sex, bandwidth, and horsepower: the more we get, the more we want.

Today, we take cloud services and our ability to get Yelp! restaurant reviews on our mobile phones for granted. But just three generations ago, the bulk of the world's computing was done by "computers" that is, humans who were facile with numbers and who were paid to do complex math problems all day, every day. The problem was that those humans were just too slow. The quest for Faster led the US military to design and build the world's first general-purpose electronic computer—ENIAC, short for Electrical Numerical Integrator and Calculator.

World War I was a conflict defined by trenches and big guns. Most of the casualties in that war were caused by artillery fire. The problem with big guns, however, has always been accuracy. Landing a 95-pound (43 kilogram) shell packed with high explosives onto a target 8 miles (13 kilometers) away requires sophisticated mathematics.[15] In the late 1930s, as World War II was simmering in Europe, the US military realized it needed to improve the mathematical tables that it used to calculate the trajectories of artillery shells.[16]

ENIAC at University of Pennsylvania, sometime between 1947 and 1955. Note the large air-conditioning vents on the ceiling. Then, as now, big computing facilities require lots of cooling. *Source*: US Army.

The only way to create those tables was with people who worked at desks and manipulated mechanical calculators like the Marchant Silent Speed, a bulky machine with some 4,000 moving parts that could handle 10-digit numbers. After Japan bombed Pearl Harbor, the US war effort went into overdrive. By 1943, when the US military launched the Manhattan Project, the need for greater computational power became even more apparent. In fact, the computational needs that came with trying to design an atomic weapon made the ones needed for the accurate firing of artillery shells look positively puny.

The need for something better was obvious to John von Neumann (b. 1903, d. 1957). Born in Budapest to a wealthy Jewish family, he began teaching at the University of Berlin in 1926. Four years later,

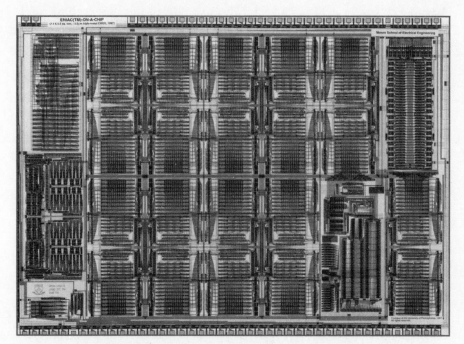

ENIAC-on-a-chip. This microchip has processing capacity equal to that of the world's first computer, ENIAC, which was completed in 1946. Fifty years later, in 1996, a team of students at the University of Pennsylvania led by Professor Jan Van der Spiegel in the Moore School of Electrical Engineering, and backed by the National Science Foundation and Atmel Corporation, replicated the architecture and basic circuitry of ENIAC. They were able to put ENIAC's capabilities onto a single chip that measured just 8 millimeters square, meaning it had an area of 64 square millimeters.[17] Its power requirements: 0.5 watts.[18] A bit of simple math shows that ENIAC-on-a-chip was about 350,000 times Smaller than ENIAC and 348,000 times more energy efficient.[19] You could fit about six of these ENIACs-on-a-chip onto a single postage stamp.[20] *Source*: Courtesy of Professor Jan Van der Spiegel, University of Pennsylvania, Moore School of Electrical Engineering.

he moved to the United States to take a position at the Institute for Advanced Study at Princeton University. He pioneered several fields of study in mathematics and was so facile with complex subjects that the nuclear physicist Hans Bethe (who won the Nobel Prize in physics in 1967) once said that he "wondered whether a brain like von Neumann's does not indicate a species superior to that of man."[21] Another colleague recalled von Neumann's photographic memory. "He was able on once reading a book or article to quote it back verbatim . . . On one occasion, I tested his ability by asking him to tell me how *A Tale of Two Cities* started, whereupon, without any pause, he immediately began to recite the first chapter. We asked him to stop after ten to fifteen minutes."[22]

As one of the principal coordinators of the Manhattan Project, von Neumann arrived at Los Alamos, New Mexico, in September 1943, ready to begin work on building the first atomic bomb. At his disposal were about twenty human computers available to do the mathematics needed. Von Neumann quickly realized that if he and his colleagues— an illustrious group of physicists, mathematicians, and scientists—were going to build an atomic bomb, they were going to need more computing power, a lot more, than what could be provided by a handful of math geeks sitting at desks.[23]

George Dyson, in his 2012 book, *Turing's Cathedral*, explains that the team at Los Alamos needed to be able to predict how an atomic bomb might behave when detonated. He writes:

> To follow the process from start to finish required modeling the initial propagation of a detonation wave through the high explosive, the transmission of the resulting shock wave through the tamper and into the fissile material (including the reflection of that shock wave as it reached the center), the propagation of another shock wave as the core exploded, the passage of that shock wave (followed by an equally violent rarefaction wave) outward through the remnants of the previous explosion and into the atmosphere, and finally the resulting blast wave's reflection if the bomb was at or near the ground.[24]

Needless to say, the mathematics involved in modeling each of those waves was daunting. Dyson said that it was not a coincidence that the atomic bomb and the electronic computer were born at "exactly the same time. Only the collective intelligence of computers could save us from the destructive powers of the weapons they had allowed us to invent."[25]

Dyson's explanation provides a quintessential example of both the dangers and benefits of innovation. The desire for Faster computing power gave humans the ability to create weapons capable of destroying the planet many times over. And yet that same computing power now provides us with the capability to create new materials, processes, and medicines.

In 1943, the US military agreed to provide the funding for ENIAC. Built at the University of Pennsylvania by a team headed by J. Presper Eckert and John W. Mauchly, the massive machine was a wonder of the pre-transistor, pre-integrated circuit era. That meant it had to rely on vacuum tubes. ENIAC contained 17,468 of them, along with 10,000 capacitors, 1,500 relays, 70,000 resistors, and 6,000 manual switches. All of those devices were connected by some five million soldered joints, most of which had to be joined by hand.[26]

In 1946, when ENIAC was completed, it weighed 27 tons, covered 240 square feet (22.3 square meters) of floor space, and required 174,000 watts (174 kilowatts) of power.[27] The computer consumed so much electricity that when it was switched on, it allegedly caused lights in the rest of Philadelphia to momentarily dim. ENIAC's enormous power demand was to become a hallmark of the Information Age. And that power demand takes us back to a familiar metric: power density.

When it was first switched on, ENIAC almost certainly had the highest areal power density of any electronic machine on the planet, about 7,800 watts per square meter. That's an astounding level of power density, particularly when compared against residential demand. An average home—and here I'm using my home in Austin, Texas, as a reference—has a power density (counting electricity only) of about 5 watts per square meter.[28] The entire city of New York, counting the five boroughs and all of its land area, including Central Park, has an areal power density of about 15 watts per square meter.[29]

The Incredible Shrinking Circuit: From the 8086 to Core i7

Top view of an Intel 8086 processor, circa 1978. *Source*: Eric Gaba, Wikimedia Commons.

Bottom view of an Intel Core i7 processor. *Source:* Wikimedia Commons.

In 1978, the year I graduated from Bishop Kelley High School in Tulsa, Intel released the 8086 processor, the brain for the first wave of personal computers. The design of the 8086 and its almost identical brother, the 8088, would become the standard for personal computer design. As one analyst put it, the 8086 "paved the way for rapid, exponential progress in computer speed, capacity and price-performance."[30] The 8086 chip sported 29,000 transistors on circuits that were 3 microns wide.[31] (A micron is one-millionth of a meter, or 1,000 nanometers.) By 2013, Intel was producing microprocessors with circuits that were just 22 nanometers wide.[32] And the company's flagship processor, the Core i7 Sandy Bridge-E, was packed with 2.27 billion transistors.[33] Thus, between 1978 and 2013, Intel increased the computing power density of its best chips 78,000-fold while shrinking its circuits more than 130-fold.

ENIAC's computing capabilities were impressive at the time. It could perform about 5,000 additions or 400 multiplications in one second.[34] It was also an enormously inefficient beast. In December 1945, during what Dyson calls a "shakedown run" of ENIAC's capabilities—a calculation for the hydrogen bomb—a pair of scientists consumed nearly one million punch cards, which were used to temporarily store the intermediate results of their calculations.[35] Needless to say, handling that many punch cards was a cumbersome process.

While von Neumann and his colleagues understood the importance of ENIAC, von Neumann knew that a Smaller Faster more powerful computer could be built. In 1946, he declared, "I am thinking about something much more important than bombs. I am thinking about computers." Von Neumann understood that speed in computation was to be sought for its own purpose. The value in Faster computers, he said, "lies not only in that one might thereby do in 10,000 times less time problems which one is now doing, or say 100 times more of them in 100 times less time—but rather in that one will be able to handle problems which are considered completely unassailable at present."[36]

Von Neumann convinced a variety of backers to provide the money for MANIAC, or Mathematical and Numerical Integrator and

This photo, likely taken in about 1952, shows mathematician and computer pioneer John von Neumann standing in front of MANIAC, the first computer to utilize random access memory. The row of cylinders to his left are the cathode-ray tubes that provided the RAM for the machine. Author George Dyson says, "The entire digital universe can be traced directly" back to the creation of MANIAC.[37] *Source*: Alan Richards, photographer. From the Shelby White and Leon Levy Archives Center, Institute for Advanced Study, Princeton, NJ.

Computer, which was built at the Institute for Advanced Study at Princeton University. (At Princeton, the machine is referred to as "the IAS Computer.") Introduced to the public in 1952, it was the first computer to utilize RAM—short for random access memory—and for that reason alone it was a milestone in the push for Faster Cheaper computing.[38]

Today, RAM is ubiquitous. The most common type of memory, it can be found in billions of electronic devices from dishwashers to cameras. But as Dyson points out, the first actual utilization of RAM was pivotal. He writes that thanks to RAM, "All hell broke loose . . . Random-access memory gave the world of machines access to the powers of numbers—and gave the world of numbers access to the powers of machines."[39] The quantity of information available in MANIAC's RAM was almost laughably small, just 5 kilobytes, less memory than is required to display a simple icon on a modern computer. But back in 1952, it was nearly miraculous.

Let's pause here for a bit of perspective: The computer I'm using to write this chapter is a MacBook Pro, which is equipped with 8 gigabytes of RAM, or about 8 million kilobytes. Therefore, my computer has 1.6 million times as much RAM as was available on MANIAC, a machine that began operating the year after my parents were married. (Walter Bryce married Ann Mahoney in 1951.)

Like ENIAC, MANIAC required lots of power. Dyson described MANIAC as being about the size of an industrial refrigerator.[40] It measured about 6 feet high, 2 feet wide, and 8 feet long, giving it a footprint of 16 square feet (1.5 square meters). But powering it required 19,500 watts (19.5 kilowatts).[41] Therefore, the areal power density of MANIAC was about 13,000 watts per square meter. That's a remarkably high power density number when you consider that modern data centers are being built for power densities of about 6,000 watts per square meter.

When we look at the history of ENIAC and MANIAC alongside that of modern data centers, we see a clear trend: the more computer power we want, the more electricity we need. As our computing needs grow, so too does power density. And few companies have chased density more avidly than Intel.

"Green" Computing Can't Power the Cloud

Facebook's initial public offering was all about superlatives. The May 2012 event was the largest-ever IPO for a US technology company and the third-largest in US history.[42] It marked, or so the hype claimed, the coming of age for social media companies. But amid the hype over the company's stock price, revenues, and growth potential, the media paid almost no attention to the vast quantities of electricity that Facebook and other tech companies need to operate their business.

In April 2012, Greenpeace spotlighted the issue of power demand in data centers in the report "How Clean Is Your Cloud?" The environmental group graded a series of technology companies, including Facebook, Apple, Amazon, and others on the percentage of what it called "dirty energy" being used by their data centers.[43] Greenpeace—which, of course, has a Facebook page—gave the social media company a "D" for what it calls "energy transparency." And the group went on to claim that it had convinced Facebook to "unfriend" coal-fired electricity.[44]

Never mind that about 40 percent of all global electricity production comes from coal. Let's consider what the "clean energy" footprint of one of these big data centers might look like.

In 2012, James Hamilton, a vice president and engineer at Amazon Web Services, wrote about Apple's new iCloud data center in Maiden, North Carolina. Hamilton was responding to Apple's claim that it was going to use solar energy to help run the site. In a blog posting called "I love solar but . . . " Hamilton calculated that each square foot of data center space would require about 362 square feet of solar panels.[45] In all, Hamilton estimated that powering Apple's 500,000-square-foot data center would require about 6.5 square miles (16.8 square kilometers) of solar panels. Hamilton said that setting aside that much space, particularly in the densely populated regions where many data centers are built, is "ridiculous on its own" and would be particularly difficult because that land couldn't have any trees or structures that could cast shadows on the panels.

Utilizing wind energy to fuel data centers would be equally problematic. To demonstrate that, consider the Facebook data center in Prineville, Oregon, which needs 28 megawatts of power.[46] The areal power density of wind energy—and it doesn't matter where you put your wind turbines—is 1 watt per square meter.[47] (I will address wind energy in a later chapter.) Therefore, just to fuel the Facebook data center with wind will require about 28 million square meters of land. That's 28 square kilometers or nearly 11 square miles—about half the size of Manhattan Island, or about eight times the size of New York City's Central Park.[48]

The mismatch between the power demands of Big Data and the renewable-energy darlings of the moment are obvious. US data centers are now consuming about 2 percent of domestic electricity. That amounts to about 86 terawatt-hours of electricity per year, or about as much as is consumed by the Czech Republic, a country with ten million residents.[49] Put another way, US data centers are consuming about 47 times as much electricity as what was produced by all the solar-energy projects in America in 2011.[50]

The hard reality is that our iPhone, Droids, laptops, and other digital devices require huge amounts of electricity. According to Jonathan Koomey, a research fellow at Stanford University who has worked on the issue of power consumption in the information technology sector for many years, data centers consume about 1.3 percent of all global electricity.[51] That quantity of electricity, about 277 terawatt-hours per year, is nearly the same as what is consumed by Mexico.[52] While that's a lot of energy, Koomey's estimate of 277 terawatt-hours doesn't account for the energy used by home computers, TVs, iPads, iPods, video monitors, routers, DVRs, and mobile phones.

In 2013, Mark Mills, a colleague at the Manhattan Institute, wrote a report called "The Cloud Begins with Coal," which put the total even higher. Mills estimated that when all the energy used for telephony, Internet, data storage, and the manufacturing of information-technology hardware is included, about 7 percent of all global electricity is being used in our effort to stay connected. That amounts to about 1,500 terawatt-hours per year, or nearly as much electricity as is used annually by Japan and Germany combined.[53]

Regardless of the precise amount of energy being used to run our digital communications network, it's readily apparent that communications-related electricity demand is growing rapidly. Between 2005 and 2010, global use of electricity in data centers grew by about 56 percent.[54] That's more than three times as fast as the growth in global electricity consumption over that same time frame.[55] It's apparent that the demand for electricity to power data centers will continue to grow as more people, and more things, get connected to the Internet. A plethora of digital devices—ranging from smart phones to GPS-enabled locators on shipping containers—is connecting to the network. In 2012, Intel estimated that there were about 2.5 billion devices connected to the Web. By 2015, it expects there will be fifteen billion Net-connected devices.[56] Ericsson predicts fifty billion by 2030.[57]

Again, the exact numbers are not as important as the trend. The push for Smaller Faster digital devices requires moving evermore information. The more computing power we use, the more electricity we consume. Big Data has always demanded Big Electron. And as we've managed to move more and more bits, we've seen a corresponding increase in the demand for electricity.

SMALLER FASTER INC.

INTEL

Official Name Intel Corporation

Website http://www.intel.com

Ownership Publicly traded, NYSE: INTC

Headquarters Santa Clara, CA

Finances Market capitalization: $119 billion[58]

2012 Revenue $53.3 billion[59]

Take a moment to look at the period at the end of this sentence. Using current manufacturing techniques, Intel Corporation could fit more than six million transistors onto that little dot.[60]

Although the vacuum tube ignited the Information Revolution, it took the transistor to make the Information Age relevant to consumers. Transistors are Smaller Faster Lighter Denser Cheaper and more reliable than vacuum tubes. And no other company has had more success at making transistors Smaller than Intel.

Of course, numerous companies compete with Intel, including Advanced Micro Devices and Texas Instruments. But it's also abundantly clear that Intel, founded in 1968, is leading the world in nanotechnology. By doing so, the company has helped make transistors ubiquitous. Imbedded in hearing aids, smart phones, TVs, dishwashers, computers, automobiles, and dozens of other items, they are among the most commonly manufactured items in the world. Intel alone produces about five billion transistors *every second*. That works out to about twenty million transistors per year for every resident of the planet.[61]

Forty Years of Smaller at Intel:
From 10,000 Nanometers to 22 Nanometers

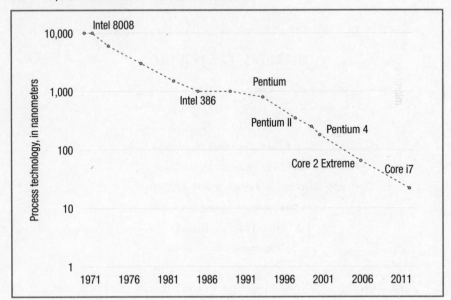

Source: Wikipedia. http://en.wikipedia.org/wiki/List_of_Intel_micro
processors.

When John Bardeen, William Shockley, and Walter Brattain
assembled the first transistor at Bell Labs in 1947, the device was
about the size of the palm of your hand.[62] Ever since then, the
transistor has been shrinking. Process technology is the manufacturing method used to put transistors and other components on
silicon chips.[63] In 1971, Intel began producing microprocessors
using a 10-micron process technology. (A micron is 1 millionth of
a meter, or 1,000 nanometers.) Forty years later, the company was
using a 22-nanometer process technology. At 22 nanometers, you
could fit more than 4,000 transistors across the width of a human
hair. By 2020, analysts believe Intel and other companies may be
utilizing a 4-nanometer process technology.[64]

From 1971 to 2011, the per-transistor price for Intel's microprocessors dropped by a factor of about 50,000. The company's
modern microprocessors run about 4,000 times as fast as the ones
used in 1971, and each transistor consumes about 5,000 times less

Forty Years of Denser at Intel:
From 2,300 Transistors per Microprocessor to 2.27 Billion

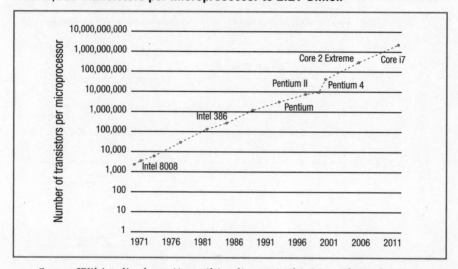

Source: Wikipedia. http://en.wikipedia.org/wiki/List_of_Intel_micro processors.

energy.[65] While it's difficult to imagine those changes, look at the graphic above, which shows how the density of Intel's microprocessors has increased over the past four decades.

In 1971, Intel was able to put 2,300 transistors on a microprocessor. By 2011, the company was installing nearly 2.3 billion on a single chip—a million-fold increase. Back in 1996, Ed Lazowska, chair of the University of Washington's Computer Science and Engineering Department, estimated that if automobile technology had advanced at the same pace as computer hardware and software, cars would be about the size of toasters and cost just $200. They'd also be able to cruise at 100,000 miles per hour while using just one gallon of fuel per 150,000 miles traveled.[66] Remember, Lazowska made that comparison in 1996.

Why pursue such high density? Obvious, says one former Intel guy: "The chip gets smaller because its transistors and wires are smaller. It gets faster because smaller transistors are faster. Smaller is also cheaper . . ."[67] The relentless push for Smaller Faster Cheaper has made Intel into a $120 billion company.[68]

Smaller Denser Cheaper: The Plummeting Cost of
Computer Storage, 1956–2010

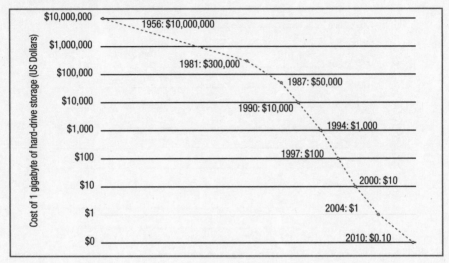

In 1956, a gigabyte of hard-drive storage cost $10 million.[69] By 2010, that same amount of storage could be purchased for about $0.10.[70] Today, thanks to Smaller Faster Lighter Denser Cheaper computers and hard drives, consumers can use up to 5 gigabytes (a gigabyte is 1 billion bytes) of online storage—in cloud services like Dropbox and Google Drive—for free. *Sources:* http://www.geekosystem.com/gigabyte-cost-over-years/; Mat Komorski, http://www.mkomo.com/cost-per-gigabyte.

10

FROM LP TO iPOD

It's a retro thing to do, no doubt about it. I recently went to Waterloo Records, Austin's landmark music store, and bought an LP—not a CD, not an MP3, but an old-fashioned 12-inch wide slab of petroleum products. My choice: the twenty-fifth anniversary edition of Paul Simon's classic album, *Graceland*. The cost, before tax: $24.99.

I didn't really need the LP. Shortly before I drove to Waterloo Records, which is about 3 miles from my house, I downloaded the same twenty-fifth anniversary edition of *Graceland* onto my computer. The download, which cost $14.99 on the iTunes store, took about twelve minutes. It came with seventeen tracks, all of the lyrics to the songs, a digital booklet, a discography of Simon's albums, and the digital copies of the music videos that Simon produced when the album came out in 1986. In fact, the digital version is, in some ways, cooler than the analog version, even though the LP came with a nice poster.

I dig records. I've had a thing for vinyl since I was a teenager and I now have about 250 LPs. CDs are cool, too. I've been buying them since about the time *Graceland* first came out. CDs are Smaller than LPs. Their sound quality is great. They are durable and can handle cold, heat, and other abuse. You can play them at home, in the car, or on the boom box. Those reasons help explain why in the first twenty-five years after CDs became commercially available back in 1982, retailers sold more than 200 billion of them.[1]

But the LP, CD—and of course, the 8-track and cassette tape—are all anachronisms. In the span of a few decades, we've gone from vinyl to CDs and from the Walkman to the iPod. Music that used to exist only when we put the needle of the turntable onto an LP spinning at 33 1/3 revolutions per minute, now lives as a digital file on a machine

1892: Thomas Edison (center) with his colleagues, gathered around his wax-recording phonograph. The machine could record about two minutes of sound. *Source*: Library of Congress, LC-USZ62–13413.

Smaller than a pack of cigarettes, or even in the cloud as a collection of bits we can retrieve at will from nearly anywhere via services like Pandora and Spotify.

That's a rather long introduction to the obvious: our music has gotten Smaller Faster Lighter Denser Cheaper. Sounds that were formerly available only in the opera house or concert hall at specific times on specific dates can now be heard nearly anywhere, anytime. The whole business of recording, buying, storing, and replaying music has become so easy and efficient that we can now purchase the soundtrack to our lives on our mobile phones without ever setting foot inside the Concertgebouw in Amsterdam or Waterloo Records in Austin.

The push toward iPods and cloud-based music led me to consider my LP collection in terms of weight and density. So I pulled out a scale

Smaller Faster Lighter Denser Cheaper Music Storage: From the LP to the iPod

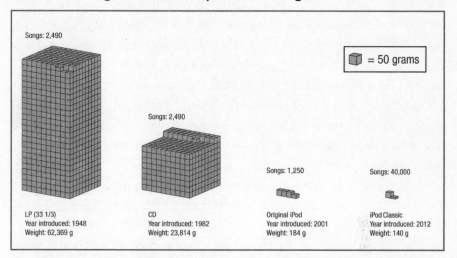

Over the span of about six decades, our music-playback systems have gotten Smaller Faster Lighter Denser Cheaper. The iPod Classic hold about 40,000 songs and weighs 140 grams—about 5 ounces—making it light enough to strap to your arm or carry in your pocket. If you attempted to carry those same 40,000 songs with you on LPs, you'd need a mule train. That quantity of music-on-LP would weigh about 1,000 kilos, or 2,200 pounds.[2] *Source*: Author calculations, Apple.

and a tape measure and tallied the results: my LPs, which hold approximately 2,500 songs, weigh about 62 kilograms and require about 95 liters of space. I then compared that to the music density on an iPod. My findings: an iPod is about 7,000 times as efficient in terms of weight as an LP. In volumetric terms, it's about 20,000 times as efficient. Put another way, if the iPod Classic—which can hold about 40,000 songs and weighs just 140 grams—had the same density as my collection of 250 LPs, it would be the size of a large refrigerator and weigh as much as a Fiat 500.

Improvements in recording have been under way ever since 1877, when Thomas Edison invented the phonograph. By 1899, when Edison began commercializing the invention for consumer use, the phonograph cost $20 (more than $500 in today's money) and could record just two minutes of sound.[3] Going from Edison's wax-cylinder phonograph to the iTunes of Steve Jobs took a century, and all along the way, sound recording and playback got progressively better. The wax cylinder gave way to 78-rpm records, which could hold about three minutes of music per side. The LP followed and increased the density of the recording fivefold. The LPs could record about fifteen minutes per vinyl side. The LP gave way to the CD, which prevailed because of its better sound quality, ease of use, and, of course, its higher density. The CD weighed less than half as much as an LP and contained six times as much music per cubic centimeter.[4]

That ability to produce music, to record sound of whatever type and play it back whenever and wherever we like is an astounding breakthrough. Imagine the expression on Thomas Edison's face if we were to hand him an iPod and a set of earbuds and then dial up a rendition of Beethoven's Ninth Symphony. There's no doubt that Edison would be gobsmacked. And he would be astounded in the same way that his fellow citizens were back in the late 1800s when he began demonstrating and selling his first recording devices.

Today, music and recording technology continues to evolve, making it Smaller Faster Lighter Denser Cheaper than ever before.

11

FROM KUBLAI KHAN
TO M-PESA

The SS *Gairsoppa* was hauling money. To be precise, it was carrying about 219 tons of silver from India to Ireland when it was torpedoed by a German U-boat on February 17, 1941. For seven decades, the wreckage of the coal-fired *Gairsoppa* and its $230 million cache of silver sat undisturbed on the bottom of the ocean in about 4,700 meters of water.[1]

In 2011, a company called Odyssey Marine Exploration found the *Gairsoppa* in a spot about 300 miles off the Irish coast.[2] Since 2011, the Florida-based company, which is publicly traded on the NASDAQ under the ticker OMEX and is excavating a number of other shipwrecks and claiming their precious-metal cargoes, has recovered more than 48 tons of silver from the sunken cargo ship.[3] The recovery of the silver from the *Gairsoppa* is one of the deepest and largest recoveries of precious metal in history.[4]

Many people dream of sunken treasure, pirate booty, and all of the romance that comes with them. But today the idea of using silver or gold to pay debts seems almost quaint. Even good old paper money is starting to appear old-fashioned. Many airlines refuse to accept cash onboard their flights. Instead, customers must pay with credit cards. While cash and credit cards continue to dominate the marketplace, in some parts of the world they are being replaced by "mobile payments," which rely on the SIM cards and data that sit inside mobile phones.

We are witnessing a move toward Smaller Faster Lighter Cheaper money. While we're in the early stages of the cash-into-digits trend, the move offers great promise because money fosters interaction. Having a reliable, trustable method of exchanging value—whether with gold, currency, or digits on your phone—builds communities and economies. Access to money is a human right. It allows people to save the fruit of their labor. It fosters the diffusion and accumulation of wealth.

Here's a brief history of money.

Humans have been using gold as a store of value for millennia. The earliest pure gold coins date to about 560 B.C. in what is now Turkey. The Bible has numerous references to gold. The element appears eight times in Genesis alone.[5] Gold coins proliferated for many reasons: the metal is shiny, durable, malleable, easy to test for authenticity, and non-reactive. We can count on it to not dissolve, mildew, or catch fire. Gold was, and still is, pretty scarce. In all of history, the total amount of gold ever mined totals about 165,000 tons, a weight equal to that of one and a half US aircraft carriers.[6] Coins made from silver (the word "coin" means "invent") have many of the same attributes as their gold cousins and have been in use for centuries. While silver and gold—atomic numbers 47 and 79, respectively—have many positive characteristics, they also have a key drawback: they're heavy.

In the thirteenth century, the Chinese emperor Kublai Khan changed the way we think about money.[7] Khan's great insight was that money—in the form of sea shells or gold coins—was valuable only if people believed in it. He also knew that the different regions of China were issuing their own coins, which made trade within his empire more difficult. So Khan created a new currency based on paper money. By decreeing that his paper money had value, his subjects believed that it did. Khan's paper money not only provided a common currency for his empire, it was also far superior to gold and silver coins for an obvious reason: it was Lighter. Being Lighter, Khan's paper money made trade Faster.

Consider, for example, the farmer rich in chickens or apples. He could, of course, transport his birds and fruit to town and barter them for something he really needs, like horseshoes or butter. But if the farmer can instead sell his products, collect some currency, and then use it to pay the farrier or the dairyman, the entire process happens Faster, with less friction for both parties. That's the point of money.

As James Surowiecki of the *New Yorker* points out in a clever 2012 article published in *IEEE Spectrum* magazine, "What matters most about money is not what it is, but what it does." Successful currencies "lubricate commerce, allow people to exchange goods and services, and thus encourage people to work and create. The German sociologist Georg Simmel described money as 'pure interaction,' and that description seems apt—when money is working as it should, it is not so much a thing as it is a process."[8]

That's an essential point: Money isn't a thing, it's a process. Money is only worthwhile because it allows us to engage in "pure interaction." We want to network. It's in our nature to do so. We want to sell, buy, haggle, argue, travel. We want to do more, of everything. We don't give a darn about the form of our money; we only care that it allows us to buy and to do, whether that means buying a bale of toilet paper at Costco or securing passage to Panama City. Having an easy method of exchanging value greases the wheels of commerce. And the slicker the lubricant, the better.

The idea of digital money is not entirely new. For decades, financial institutions, corporations, and individuals have relied on wire transfers to exchange money. In 2012, according to the Fedwire Funds Service, the wire-payments network operated by the US Federal Reserve, nearly $2.4 trillion per day was moved by wire transfer. (For reference, that amount of money is approximately equal to the GDP of the United Kingdom.)[9] Furthermore, the volumes of money being moved electronically continues to grow. Between 1987 and 2012, the amount of money moved through the Fedwire system has nearly quadrupled.[10] For banks and big companies, digital money has long been a fact of life. That hasn't been the case for consumers. Sure, lots of people forgo cash by carrying credit cards and debit cards, which are a form of electronic payment. But those methods are not truly digital.

Perhaps the most remarkable thing about the move toward digital money and mobile payments is that the trend shows its strongest growth in Africa, a continent that has long been seen as a laggard in nearly every other type of development. In countries like Kenya and South Africa, pure interactions are happening with currency that weighs nothing at all. Instead, it exists only as digits on phones. Currency is being exchanged on the simplest cell phones using the simplest technology: text messaging, or SMS, for short message service. And that in itself is amazing.

In 2011, some 5.9 trillion text messages were sent. By 2016, that number is expected to rise to 9.4 trillion. As the savvy South African journalist Toby Shapshak has noted, mobile phones have "become the most-used devices in the world." While teenagers and college students in the United States are mesmerized by Instagram, Snapchat, and other iPhone flotsam and jetsam, Shapshak points out that in the developing world, "the vast majority of people still use them [their phones] for their primary functions: voice calls and text messaging. SMS is still the king of communication. Long may it reign."[11]

SMS is the perfect communication system for Africa, where the majority of consumers can't afford high-dollar devices like the iPhone or an Android-powered device. Instead, as Shapshak points out, they use older Nokia phones or the decade-old Samsung E250, a phone that has been dubbed the AK-47 of African telecom because it is cheap and nearly indestructible. While the majority of African consumers lack flashy phones, they are getting lots of leverage from them. Most people living in sub-Saharan Africa are "unbanked," meaning they don't have bank accounts or credit cards. But nearly all of them have a phone. And those phones are giving millions of Africans access to a trustworthy, secure, money ecosystem. Cellular phone companies and banks are providing digital cash to millions of people who've never stepped inside a bank or managed a cash register. It's almost as though Kublai Khan gave millions of Africans a single currency and told them to start trading.

Khan was a clever ruler, but he understood that making and managing cash costs money. That's still true today. Hiring the British company De La Rue or one of the other specialty firms who print currencies is expensive. Producing a single banknote costs about four cents.[12] That adds up if you need tens of millions of them. Banknotes are vulnerable to counterfeiters and worse yet, they wear out after a

Paper money may be more convenient than carrying bags of gold or silver, but that paper is only valuable if people believe in it. These banknotes were printed by the Confederate States of America between 1861 and 1864. Each note was hand signed and numbered. During the Civil War, in an effort to upset the economy of the Southern states, the Union began counterfeiting Confederate currency and distributing it in the South. The flood of money fueled inflation and helped undermine confidence in the currency, which had become worthless by the time the war ended in 1865.[13] For the first few decades of the twentieth century, all of the world's biggest economies were on the gold standard. If you wanted to trade your paper money for gold, you could be assured of a set amount of gold in return.[14] Today, no country backs its currency with gold. Instead, the value of each country's currency "floats" in relation to other currencies. There has been a decades-long argument as to whether this "fiat" money system—that is, a system in which the currency is not linked to any specific asset—can last for the long haul. Gold may be heavy and difficult to move around, but people have been believing in it for 2,500 years.[15] *Source:* Library of Congress, LC-USZ62–110272.

few years. Even if a poverty-stricken country like Democratic Republic of the Congo or Niger wanted to print a big batch of currency to help their economies, doing so would create other costs. Large quantities of cash require heavy trucks, armed guards, and safes.

The physical costs of managing currency go far beyond Africa. During the early months of the Second Iraq War, about 150 tons of US currency worth some $12 billion was sent in several air shipments to military bases near Baghdad. In 2004, in the largest single shipment, the US Federal Reserve sent $2.4 billion in $100 bills (weighing 30 tons) to the ancient city. The greenbacks were needed so that the US military and the new Iraqi government could contract for services, hire personnel, and get the war-torn economy up and running again.[16]

Or consider the challenge of managing currency in China. The Chinese government refuses to print any bills larger than the 100-*renminbi* note, which carries the portrait of Mao Zedong. But the 100-renminbi note (the renminbi is also known as the yuan) is only worth about $16. In the United States, the largest bill is the $100 note. In the European Union, it's the 500-Euro note, which is worth about $650. Relatively few Chinese use credit cards or checks. Thanks to Chinese citizens' distrust of their government—and the government's distrust of its citizens—many Chinese deal strictly in cash. But because China's currency holds relatively little value, consumers have to carry lots of bills, particularly when making large transactions, such as buying a car. That means that the Chinese government has to spend a lot of money keeping the Chinese economy stocked with paper currency. In fact, the business of currency in China requires a vast industrial enterprise.

In 2013, David Barboza, a reporter for the *New York Times,* reported that the China Banknote Printing and Minting Corporation "runs 80 production lines with 30,000 workers, six bank note companies, two paper mills, a printmaking company, a plate-making corporation, and a firm that produces special anticounterfeiting security lines."[17] Barboza estimated that when adjusted for the size of its economy, China has about five times as much currency in circulation as does the United States. In all, about 40 percent of all the paper money on the planet is in China, even though China's GDP of about $8.2 trillion accounts for only about 11 percent of global GDP, which totaled about $72 trillion in 2012.[18]

Cash will certainly persist for years to come in markets all over the globe. But when money lives inside your phone, there's no need for printers, paper mills, or safes. Just as Africa leapfrogged the idea of landline telephones and went straight to digital mobile phones, so, too, is Africa vaulting over the idea of currencies and going straight to digital money.

In the span of 70 years, we've gone from shipping silver on a coal-fired steamship to M-PESA, a company that is transacting millions of dollars in commerce every day on a device that weighs as little as 85 grams (3 ounces).[19] The most important part of that phone—the SIM card that stores the personal data—weighs less than half a gram.[20] Money has gotten Smaller Faster Lighter Denser Cheaper, astoundingly so. And it's done so in a blindingly short amount of time. After millennia—or what seems like it—of fumbling around for nickels and dimes amid the couch cushions, we now have virtual money. For lots of transactions, that digital dough's far superior to coins and banknotes.

This qualifies as a Big Idea. The world's biggest companies are jumping into the digita-money/mobile-payments game. Chase, Walmart, Google, AT&T, American Express, MasterCard, and dozens of other companies are getting into mobile payments, all trying to, in effect, coin their own money. They are chasing, as *Fortune* magazine put it, "millions of merchants, billions of transactions, and trillions of dollars in commerce."[21]

In early 2013, the coffee giant Starbucks said it was handling more than three million mobile payment transactions per week, all of them in the United States.[22] By 2016, the consulting firm Gartner Inc. expects that about 450 million people around the world will be using mobile payments to conduct more than $600 billion worth of transactions. If that occurs, it will be a nearly fourfold increase over the $171 billion in mobile transactions that occurred in 2012.[23] Consumers in India, Bangladesh, Pakistan, Japan, China, Canada, Australia, and other countries are rapidly adapting to mobile payments.[24] But for now, Africa remains the dominant player.[25] By 2013, about 80 percent of the world's mobile payment transactions were happening in East Africa.[26] And the biggest player in East Africa is M-PESA.

SMALLER FASTER INC.

SAFARICOM

Official Name Safaricom, Ltd.

Website http://www.safaricom.co.ke/ (English)

Ownership Public (Ticker: SAFCOM, Nairobi Stock Exchange)

Headquarters Nairobi, Kenya

Finances Market capitalization: $4.3 billion[27]

Annual Revenue $1.25 billion[28] (2012)

M-PESA was launched in March 2007 by Safaricom, a Kenyan mobile phone provider that is 40 percent owned by mobile giant Vodafone.[29] Within 16 months, M-PESA—the "m" stands for mobile, while "pesa" is the Swahili word for money—had 3.6 million customers, and the system was adding 10,000 new registrations every day. By July 2008, the system was handling 21 billion Kenyan shillings ($245 million) of transactions per month, with an average value of 2,800 Kenyan shillings, (about $33) each.[30] The system is simple: customers who have cash in their pockets can go to any of M-PESA's agents and have that paper money converted into mobile money. They can also reverse that process.

M-PESA has grown rapidly thanks to Safaricom's dominance of the Kenyan mobile phone sector. By 2013, Safaricom had nearly as many subscribers, about 19 million, as Kenya has adults. And of those 19 million phone subscribers, about 15 million were using M-PESA.[31] (Kenya's population is about 43 million.) Those 15 million have been using M-PESA to pay for everything from utilities and insurance to school fees and health care. They can also transfer money directly to another person. The system is easy to use: when an M-PESA user wants to buy something from a vendor, he uses his phone to transfer the required amount via text message.

While mobile payments and digital money have many advantages over coins and currency, we must also acknowledge the risks. Cash has been king for a long time, and that royal status brings with it a measure of privacy. Cash buys anonymity. For illicit transactions—drugs and prostitution to name just two—cash helps assure that the buyer and seller can't be easily tracked. Cash deals can also help sellers avoid the tax man. Cash is also handy for perfectly legitimate transactions. A plumber, Web designer, or doctor may prefer to be paid in cash because it's Faster and easier than dealing with checks or credit cards. Cash talks to the cabbie, the waiter, and the maître d'hôtel in a way that a promised payment via mobile phone cannot. A $100 bill might be cumbersome, but it cannot be "declined" because a computer network is down or because of a faulty credit score.

The potential loss of anonymity that could accompany the phaseout of cash has political overtones. In 2012, Jerry Brito, a research fellow at the Mercatus Center at George Mason University, wrote that in a cashless world, it would be "easier for governments and corporations to spy on our transactions" and they could "gain greater control over which transactions to allow at all." Brito points to the case of WikiLeaks. After the renegade media group released a trove of embarrassing documents, payment processors like PayPal and Visa were pressured to quit doing business with it. Rather than fight the US government, the companies agreed and WikiLeaks has been struggling to stay afloat ever since. "Imagine if the only way to support unpopular causes was with easily controlled e-money. Certain transactions could be disallowed by law, political pressure or corporate fiat, and anonymous giving would be impossible," says Brito. "One could not make a purchase at a gay bookstore or a pregnancy clinic without knowing that some-where there's a permanent record of the transaction. And there might not be any transaction that couldn't be subpoenaed in a divorce or other legal proceeding."[32]

While cash provides anonymity, some online payment outfits are already providing the capability to move money across inter-national borders with little governmental oversight. That became apparent in mid-2013, when the operators of Liberty Reserve, an online payment company, were indicted by the US attorney in Manhattan. The indictment claimed that Liberty Reserve was a front for international money laundering and that the company had laundered some $6 billion. But other online payment systems like Russia-based WebMoney and Panama-based Perfect Money are, as of this writing, still operating. Law enforcement officials claim that the sites have become havens for criminals who are laundering money from child pornography, the drug trade, weap-ons trafficking, and human trafficking.[33]

Although digital money can be used by criminals, it can also be used to fight corruption. In 2012, Jessica Leber wrote an ar-ticle for MIT *Technology Review* in which she told about a group of Afghan policemen in Wardak province who began getting their wages paid through their cell phones. The payments came through M-Paisa, a mobile payment system run by Afghanistan's biggest telecom company, Roshan, which was modeled on Kenya's M-PESA. In 2009, immediately after the first pay period in which they got paid through their phones, the policemen assumed they'd gotten a raise. The reality was that for the first time, their wages weren't paid in cash, and therefore, weren't subject to skimming by their superior officers, who had been stealing about 30 percent of the money.[34]

Leber pointed out that about half of the 700,000 government employees in Afghanistan don't have bank accounts. And getting cash to those employees is fraught with danger because of the country's security problems. Paying them with digital cash on their phones could help alleviate both security and corruption issues. The scale of the latter problem is both staggering and depressing. In 2012, Afghans paid nearly $4 billion in bribes, an amount that's roughly double the country's domestic tax revenue.[35] The corrosive effect of corruption is common in other developing countries as well. In India, the consulting firm McKinsey has estimated that tens of billions of dollars are "leaking" out of the Indian economy every year due to the pilfering of welfare and government payments by insiders.[36] In late 2013, Michael Joseph, the former CEO of Safaricom, told the *New York Times* that India is "probably the most exciting market for mobile money in the world." By 2015, some analysts are expecting mobile payments in India could total $350 billion per year.[37]

Mobile payments are not going to cure the world's corruption problems or bring all of the people who are living in poverty into prosperity. But what has happened with M-PESA and other mobile payment schemes provides a window into what can happen if people who don't have access to paper money or formal banking systems are allowed to engage in commerce with digital currency. Back in 1976, the economist Fredrich Hayek argued for the deployment of lots of currencies: "We have always had bad money because private enterprise was not permitted to give us a better one."[38]

It remains to be seen what will happen with multinational currency schemes like Bitcoin, a "cryptocurrency" that can be transferred through a computer or smart phone without relying on a financial institution. Bitcoin calls itself the world's first decentralized digital currency, and it has launched a flurry of financial speculation as investors have piled into Bitcoins with the hope that they will gain in value in the years ahead. It's far too soon to predict what will happen with Bitcoins. It's also too soon to know whether emerging "peer-to-peer" payment applications like Venmo, which allows friends to pay each other small sums of money, will endure over the long haul.[39]

Nevertheless, it's obvious that M-PESA and the other modes of digital payment are here to stay, and that is a positive development.

Money—whether it lives on your phone or in your wallet—is the oxygen that allows the economy to breathe. Money allows us pure interaction. By making money Smaller Faster Lighter Cheaper, Safaricom and its many competitors in the mobile payments business are showing the way toward a brighter future, one that Kublai Khan would surely understand.

12

DENSITY AND THE
WEALTH OF CITIES

We humans are unusual in the animal kingdom in that we are eusocial. Only a handful of species—including bees, ants, termites, and us—like to live in large colonies.[1] For millennia, our innate eusociality, our desire to divide the chores among a large group of individuals and co-operate for the common good, has been driving us to cities.

Cities, said Descartes, provide "an inventory of the possible." And for about 7,000 years, we have been escaping the cloister of isolated farms and organizing ourselves into colonies not of dozens of people, or hundreds, but of thousands. By about 3000 B.C., the city of Ur in Mesopotamia likely had about 24,000 people.[2] Today, the world's largest cities have a thousand times as many residents. Tokyo alone has some 37 million people.[3] Cities have been built all over the planet, from Cape Town, on the southern tip of Africa, to Anchorage, which sits about 400 miles from the Arctic Circle.

Our future lies in density. That's true for all kinds of human activities, from energy and food production to computing and communications. And few trends are more illustrative of our quest for density than the never-ending procession of people into cities.

As more people move into cities, we are moving more closely together. In 1950, the world had a population density of about nineteen people per square kilometer. Today, we have about fifty-three people per square kilometer, and by 2100 we may have as many as seventy people per square kilometer.[4] Of course, those numbers reflect the world's growing population, but it's equally obvious that people are moving to cities because that's where the opportunities are.

Market Street in San Francisco before the earthquake of 1906. Note the tangle of electric wires and poles in the foreground. *Source:* Library of Congress: LC–USZ62–98494.

In his 2011 book, *Triumph of the City,* Harvard economics professor Edward Glaeser wrote that "the world isn't flat; it's paved." Cities "have been the engines of innovation since Plato and Socrates bickered in an Athenian marketplace." What allows that innovation? Density. "Cities are expanding enormously because urban density provides the clearest path from poverty to prosperity."[5]

Cities mean density. Density means prosperity. Density means innovation. A few years ago, the theoretical physicist Geoffrey West did an analysis of cities and innovation, and by looking at inventions and patents, he found that a city that was 10 times as large as its neighbor wasn't just 10 times more innovative than the Smaller one; it was 17 times more innovative. And a city with 50 times the population of a small town was 130 times more innovative.[6]

West's findings—that higher population density results in more innovation—were corroborated in 2005 by the Philadelphia Federal Reserve. In a report on the "inventive output of cities" the authors, led by economist Gerald Carlino, found that "the nation's densest locations play an important role in creating the flow of ideas that generate innovation and growth." The report concluded that the number of patents per capita is "20 percent higher in a metropolitan area with an employment density (jobs per square mile) twice that of another metropolitan area." The report also concluded that there is an "optimal employment density" for maximizing patent output. That number: 2,150 jobs per square mile, or roughly the level of Philadelphia or Baltimore.[7]

There is an even simpler way to describe the conclusions reached by West and Carlino. Matt Ridley, author of *The Rational Optimist* and several other books, says cities are where "ideas go to have sex."[8]

Ideas will keep shagging, and innovation will thrive because people keep moving into cities. In his 2009 book, *Whole Earth Discipline*, environmentalist, publisher, and author Stewart Brand writes, "In 1800 the world was 3 percent urban; in 1900, 14 percent urban; in 2007, 50 percent urban. The world's population crossed that threshold—from a rural majority to an urban majority—at a sprint. We are now a city planet." Brand continues, "Every week there are 1.3 million new people in cities. That's 70 million a year, decade after decade. It is the largest movement of people in history."[9] Brand adds that "cities are probably the greenest things that humans do."[10]

Cities are green because each city resident generally requires less stuff—concrete, steel, glass, gasoline—than their suburban counterparts. Not only are cities green, they are also powering the global economy. In 2012, the McKinsey Global Institute estimated that through 2025, about 65 percent of all global economic growth will occur in cities.[11] It also predicted that by 2025, urban consumers are "likely to inject about $20 trillion a year in additional spending into the world economy."[12] And if there was any doubt about the outsized economic impact of cities, the McKinsey study says that the world's top six hundred cities, which are home to about 20 percent of the world's population, account for about $34 trillion, or more than half, of all global GDP.[13]

Bruce Katz and Jennifer Bradley of the Brookings Institution, estimate that in the United States, the one hundred largest metropolitan areas contain about two-thirds of the population but generate 74 percent of America's GDP. "In fact," they wrote in 2011, "metro areas generate the majority of economic output in 47 of the 50 states, including such 'rural' states as Nebraska, Iowa, Kansas, and Arkansas."[14]

The triumph of the cities—to borrow Glaeser's title—comes from their ability to raise the living standards of their inhabitants. There is a clear correlation between population density and prosperity. The world's most populous regions also tend to be the most prosperous. Sure, we have the wealth of nations (as Adam Smith duly noted back in 1776), but what we really have is the wealth of cities. Sure, we have lots of countries, but we are really a world of cities. We don't visit Brazil or China; we visit Rio and São Paolo, Shanghai and Beijing.

World Bank data show that highly urbanized countries are, on average, richer than ones that aren't. This makes sense. As countries industrialize and their economies expand into new sectors, they develop new technologies. This process is usually centered in cities because that's where the workers are. In turn, people flock to cities for the new opportunities created by industrialization. They want to learn new skills, and, in the process, make more dough than their rural cousins. The United States, United Kingdom, Sweden, and other developed countries exemplify this trend. By contrast, countries with low population densities tend to miss out on the benefits of such economic growth.

While cities offer higher incomes, better restaurants, and more culture, they can also be crowded, noisy, and in some cases, plagued by squalor. Cities are also targets for terrorism. And while terrorism remains an infinitesimally small risk for the average person, the bombings at the Boston Marathon in 2013, the al-Qaeda attacks on the World Trade Center in New York in 2001, and the sarin gas attack on the Tokyo subway in 1995 are all examples of terrorists targeting city dwellers. Cities are never perfect, but as demographer Joel Kotkin writes in his 2005 *The City: A Global History*, "Humankind's greatest creation has

Denser Means Richer: Highly Urbanized Countries Are Wealthier

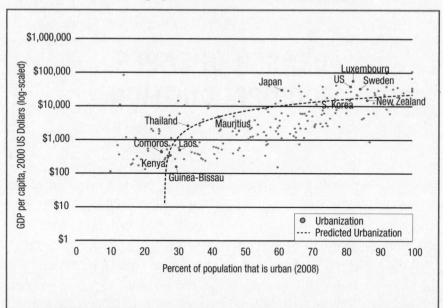

This graphic shows the strong correlation between rates of urbanization (and therefore, population density) and GDP-per-capita. The economic powerhouses of the world, countries like Japan and the United States, are clear examples. However, this even holds true in the case of a "transitioning" country like Mauritius. An island nation off the southeast coast of Africa, Mauritius is widely considered one of the most successful African countries, both economically and politically. Not coincidentally, it also has one of Africa's most urbanized populations. *Source*: World Bank Development Indicators Database.

always been its cities. They represent the ultimate handiwork of our imagination as a species, testifying to our ability to reshape the natural environment in the most profound and lasting ways."[15]

But we must also acknowledge that our ultimate handiwork, the city, has been made possible by the farm and our ability to get food products from the farm to consumers Faster than ever before. Faster Cheaper food makes cities possible. And Cheaper food has been the result of innovation that is allowing Denser food production.

13

DENSER CHEAPER FOOD PRODUCTION

In 1968, Paul Ehrlich grimly declared, "The battle to feed all of humanity is over. In the 1970s hundreds of millions of people will starve to death in spite of any crash programs embarked upon now. At this late date nothing can prevent a substantial increase in the world death rate."[1]

When Ehrlich made his dire prediction, the global population was 3.5 billion.[2] Today, the world has more than twice that many people (about seven billion), and yet the death rate today is lower than it was when Ehrlich made his claim.

What happened? Better farming happened.

What Ehrlich overlooked, and his fellow catastrophists continue to overlook, is our capacity for innovation, and in this particular case our ability to produce more food with less land. The result of that Denser food production: Cheaper food. In 2013, Keith Fuglie, an analyst for the US Department of Agriculture's Economic Research Service pointed out that in the advanced economies of the world, "people now spend 15 percent or less of their disposable income on food. It has never been lower." Fuglie further points out that Thomas Malthus published his famous "An Essay on the Principle of Population," in 1798.[3] Just nine years earlier, in 1789, on the eve of the French Revolution, "it took nearly the entire daily wage of an unskilled worker to buy two loaves of bread, enough to feed a family of four. Today it takes a Parisien about 15 minutes working at minimum wage to do the same."

Denser farm production—meaning the amount of food (or fiber) that can be produced from a given area of land—can easily be seen in the numbers. Since 1950, the food supply per capita has increased by about 30 percent, even though the amount of land per capita has fallen by about half.[4] In 1950, the world had about 2.5 billion people. Today, we have seven billion. Thanks to better agricultural production methods, we have been able to dramatically increase food production while adding only small amounts of new cropland. Hybrid seeds, more powerful farm equipment, wider use of fertilizers, and other technologies have allowed the world's farmers to dramatically increase the volume of food they can produce from their acreage.

There's no question that organic food has surged in popularity in recent years. Grocers like Whole Foods Market (market capitalization about $23 billion) that sell organic products are seeing huge increases in their market share.[5] Meanwhile, industry groups like the Organic Trade Association point out that sales of organic food and beverages are soaring. Between 1990 and 2010, in the United States alone, sales of organic products rose from $1 billion to nearly $27 billion.[6] The US Department of Agriculture has claimed that organic agriculture is "one of the fastest growing segments" of domestic farming.

All that may be true, but it doesn't mean that organic production is the best way forward. Many studies have shown that organic agriculture lags far behind conventional farming when it comes to productivity. Among the most prominent is a report by Verena Seufert, Navin Ramankutty, and Jonathan Foley that was published in *Nature* in May 2012. The report examined published literature on yields from farms around the world. While the yields depend on geography and farming methods employed, the authors concluded that "organic yields are typically lower than conventional yields." They found that the results "range from 5 percent lower organic yields (rain-fed legumes and perennials on weak-acidic to weak-alkaline soils), 13 percent lower yields (when best organic practices are used), to 34 percent lower yields (when the conventional and organic systems are most comparable)."[7]

James E. McWilliams, a history professor at Texas State University, has written extensively about food production. A vocal advocate for animal rights and veganism, he's also a prolific and fearless debunker of the hype over organic food. In a March 10, 2011, essay in *Slate*, he pointed out that global population is likely to increase by some 2.3 billion over the next four decades. That increasing population, combined with the emerging middle class in developing countries like China and India, will require the world's farmers to grow "at least 70 percent more food than we now produce." Add in the latest projections from the UN's Food and Agriculture Organization, which show that there's very little arable land left to expand production, and the conclusions, says McWilliams, should be obvious to everyone. "Barring a radical rejection of the Western diet, skyrocketing demand for food will have to be met by increasing production on pre-existing acreage," he wrote. "No matter how effectively we streamline access to existing food supplies, 90 percent of the additional calories required by midcentury will have to come through higher yields per acre."[8]

McWilliams's point is essential: farmers must be able to produce more food without increasing the size of their farms. In other words, the density of their production must increase. To bolster his thesis, McWilliams points to a 2011 analysis of US Department of Agriculture data that was done by Steve Savage, a San Diego–based plant pathologist who has more than thirty years of experience in agricultural technology.

Savage found that organic farming produces about 29 percent less corn and 38 percent less winter wheat than the same acreage that is conventionally farmed. In his summary, Savage concluded that if the United States had relied on organic agriculture to match the full output of all US crops in 2008, "it would have been necessary to harvest from an additional 121.7 million acres of cropland . . . That additional area would represent a 39 percent increase over current US cropland." To put that in perspective, Savage points out that the additional cropland needed for organic production would be about 190,101 square miles (492,363 square kilometers), which would be about the size of all the "current cropland acres in Iowa, Illinois, North Dakota, Florida, Kansas, [and] Minnesota combined." He adds that on a land-area basis that much additional territory would be nearly as large as Spain, or about three-quarters the size of Texas.[9]

Given these facts, a full-scale transition to organic production makes no sense at all. In fact, any wide-scale effort to enforce agricultural techniques that will decrease the density of production could be a recipe

Denser Farming: Global Grain Production Is Keeping Pace with Population Growth

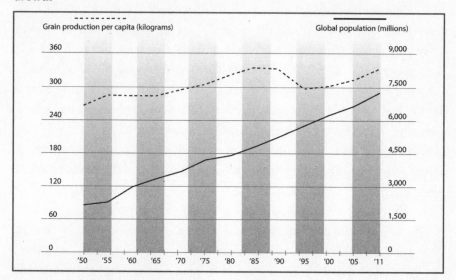

Source: Earth Policy Institute.[10]

for higher prices, increased deforestation, and possibly even mass starvation. Better farming methods not only mean Cheaper and more abundant food for people living in cities, they also mean better livelihoods for farmers. As the International Food Policy Research Institute noted in a recent report on poverty reduction efforts, "agricultural growth has a high poverty reduction payoff . . . A one percent per annum increase in agricultural growth, on average, leads to a 2.7 percent increase in the income" of the poorest populations in developing countries.[11]

Between 1950 and 2011, the world's farmers tripled the amount of grain they produced per hectare of land under cultivation.[12] But some influential environmentalists continue to advocate policies that will have the opposite effect. They have been particularly vociferous when the subject is genetically modified organisms (GMOs). For instance, Greenpeace is adamantly opposed to genetically engineered crops. The group says that "GMOs should not be released into the environment since there is not an adequate scientific understanding of their impact on the environment and human health."[13]

Greenpeace has made that claim even though numerous studies have found that GMOs are safe. In 2012, the American Association for the Advancement of Science concluded that foods containing GMOs are "no riskier than consuming the same foods containing ingredients from crop plants modified by conventional plant improvement techniques." The report goes on: "The World Health Organization, the American Medical Association, the US National Academy of Sciences, the British Royal Society, and every other respected organization that has examined the evidence has come to the same conclusion."[14]

Not only are GMOs safe to eat and produce, they could also provide a big boost to nutrition. Thanks to work done by Swiss plant biologist Ingo Potrykus, the world now has Golden Rice, a type of rice that has been altered by the introduction of genes that express beta-carotene, a substance that when ingested, is converted by the body into vitamin A, which plays a critical role in vision, the immune system, and bone growth.[15] The World Health Organization estimates that some 250 million preschool children around the world are deficient in vitamin A and that as many as 500,000 children per year become blind due to vitamin A deficiency.[16]

But don't bother Greenpeace with these facts. In 2012, the environmental group declared that Golden Rice is "environmentally irresponsible, poses risks to human health, and compromises food security."[17]

In summary, Greenpeace is opposed to GMOs like Golden Rice, even though they are more nutritious, help reduce fertilizer use, and help increase yields and therefore preserve forests. They can also help feed the poor. And yet, Greenpeace claims that it is defending "the natural world."[18]

Although it has taken more than a decade, a decade marked by controversy and anti-GMO campaigning, Golden Rice is finally making it into production. The first crops of the new strain were planted in the Philippines in 2013.[19]

Whether the technology is GMOs, more effective use of fertilizer, or more precise planting, it is readily apparent that Denser food production is a net positive for both humans and the environment. In 2012, Jesse Ausubel, the head of the Program for the Human Environment at

Rockefeller University—along with two coauthors, Paul Waggoner and Iddo Wernick—published a remarkable paper that predicted a "peaking in the use of farmland." Thanks to better technologies, over the next five decades "humanity is likely to release at least 146 million hectares" of farmland, an area that would be one and a half times the size of Egypt, two and a half times the size of France, or about ten Iowas. Ausubel and his coauthors call this reduction in the need for farmland "sparing land for nature." They concluded their paper by writing that in fifty years or so, the Green Revolution, "may be recalled not only for the global diffusion of high-yield cultivation practices for many crops, but as the herald of peak farmland and the restoration of vast acreages of nature . . . We are confident that we stand on the peak of crop land use, gazing at a wide expanse of land that will be spared for nature."[20]

14

THE FASTER THE BITS, THE FREER THE PEOPLE

A decade or so ago, Mohammed Bouazizi's self-immolation in the small Tunisian town of Sidi Bouzid likely would have gone unnoticed by anyone who didn't actually live in that dusty little town. Instead, Bouazizi's suicide in 2010 and, more particularly, the rapid dissemination of images of the protests in Tunisia that followed it, became the "big bang" that ignited the Arab Spring. The turmoil that has resulted continues to this day in Libya, Egypt, Syria, and other countries. The collapse of those Arab regimes is largely the result of our ever-more-connected world. In the age of text messages, Twitter, YouTube, Facebook, and Google, bits of information simply cannot be controlled forever. Bits want to travel, and the Faster they flow, the more we want them to flow.

Thanks to smart phones, Gutenberg's press now rides in our pockets. The ability to instantly publish photos, video, and other documents bodes well for freedom-minded people everywhere. If there's one indisputable truth about the Information Age, it's this: the Faster the bits, the freer and richer the people. The inverse of that statement is also true: the slower the bits, the less free and poorer the people.

As mobile phones, the Internet, and other communication technologies have proliferated, more and more people are able to access, and distribute, more information. The people who live in countries where information is allowed to travel fast and freely are the same ones who can get wealthier and have more liberty.

Meanwhile, the countries that are slowing, or even stopping the bits, are impeding innovation and slowing, or even stopping, their economies. The countries that restrict access to mobile phones, filter or restrict access to the Internet, censor journalists, and restrict academic freedom are imprisoning their people, making themselves poorer and falling further and further behind the rest of the world.

To prove this point, look at Netindex.com, a Web site that lists the places with the fastest broadband speeds. The top ten includes some of the wealthiest locales on earth, including Hong Kong, Singapore, Luxembourg, Japan, and Sweden.[1] Moving down the list in broadband speed shows that broadband connectivity almost works as a proxy for per-capita GDP.[2]

James Fallows, national correspondent for the *Atlantic*, illustrated that point in a 2012 essay published in the *New York Times*. He wrote that China's ongoing efforts to control the flow of information on the Internet is emerging as a major hindrance to that country's growth. In the United States, Fallows points out that the Internet can be slow due to infrastructure constraints. "In China, it's slow because of political control: censorship and the 'Great Firewall' bog down everything and make much of the online universe impossible to reach. 'What country ever rode to pre-eminence by fighting the reigning technology of the time?' a friend asked while I was in China last year. 'Did the Brits ban steam?'"[3]

Fallows's essay underscores the critical difference between the countries that allow bits to move fast and those that try to slow them down. The countries that are succeeding in the modern world are the ones that are facilitating the free flow of people, goods, and ideas. They are promoting Faster connectivity, journalistic independence, and academic freedom. The free flow of information is essential to the process of innovation, which builds wealth.

The close correlation between restricted information flows (slower bits) and poorer people can be seen by looking at North Korea, Cuba, and Iran, all three of which are listed as "authoritarian regimes" in the Democracy Index compiled by the Economist Intelligence Unit.[4] Those same three countries were also included in the 2012 list of the "Ten Most Censored Countries" compiled by the Committee to Protect Journalists. The CPJ points out that countries that censor the media have certain commonalities, including "some form of authoritarian rule.

Their leaders are in power by dint of monarchy, family dynasty, coup d'état, rigged election, or some combination thereof."

The CPJ also noted the connection between free-flowing information and wealth. "Lagging economic development is another notable trend among heavily censored nations. Of the ten most censored countries, all but two have per capita income around half, or well below half, of global per capita income . . . The two exceptions are Saudi Arabia and Equatorial Guinea, where oil revenues lead to much higher per capita income than the global level. But both of those countries are beset by vast economic inequities between leaders and citizens."[5]

All of the countries on the CPJ's list work to slow, or even halt, the flow of information to their people. The most obvious example of this is North Korea, a country ruled by what may be the most secretive and despotic regime on the planet. North Korea has a population of about 25 million people but only about one million of them have mobile phones.[6] The per-capita GDP of North Korea is $1,800.[7] That's a fraction of the global average per-capita GDP, which in 2012 was $12,500.[8]

Or consider Cuba, the country that has been run with an iron fist by the Castro brothers, Fidel and Raul, since 1959. Although Cuba has 11.2 million people, only about 10 percent of them have access to mobile phones. In addition, access to the Internet is tightly controlled by the government. In 2011, some 2.6 million Cubans had access to the Internet, but as Reuters notes, almost all of those users "were likely on the local intranet through government-run computer clubs, schools and offices." Ordinary Cubans cannot access the World Wide Web without government permission.[9] The per-capita GDP of Cuba: $10,200, again, lower than the average for the world.

Or consider Iran, which ranks 159th in the Democracy Index out of 167 countries on the list. The Iranian regime, according to the Committee to Protect Journalists, "uses mass imprisonment of journalists as a means of silencing dissent and quashing critical news coverage." In addition, "Iranian authorities maintain one of the world's toughest Internet censorship regimes, blocking millions of websites, including news and social networking sites."[10] The per-capita GDP of Iran:

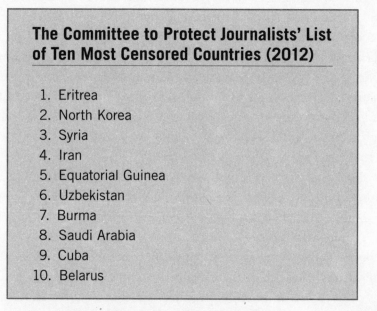

The Committee to Protect Journalists' List of Ten Most Censored Countries (2012)

1. Eritrea
2. North Korea
3. Syria
4. Iran
5. Equatorial Guinea
6. Uzbekistan
7. Burma
8. Saudi Arabia
9. Cuba
10. Belarus

Source: The Committee to Protect Journalists.

$13,300. That number is slightly above the world average, but it also reflects Iran's oil and gas resources.

In 2012, the theocrats who control Iran launched an effort to create a "Halal" Internet. ("Halal," which means "lawful" in Arabic, refers to the things that are permissible under Islamic law.) According to the Electronic Frontier Foundation, the Iranian ministry of telecommunications decreed that a number of domestic institutions, "including banks, telecom companies, insurance firms, and universities are now prohibited from dealing with emails that do not come from an 'ir' domain name. This could mean that customers who use foreign email clients such as Gmail, Yahoo!, and Hotmail will have to switch to domestic Iranian accounts, which are subject to Iranian legal jurisdiction."[11] In other words, the Iranian regime isn't content with merely blocking Web sites and prohibiting access to the World Wide Web; it also is insisting that the country's biggest institutions use only the "ir" domain for e-mail, which will then make it easier for the state to surveil its citizens' communications.

Comparing the wealth of countries that have Faster flows of information with those who actively try to slow information flows requires us to combine data sets from several organizations. In addition to the CPJ's list of most-censored countries, there are the OpenNet Initiative (ONI), which tracks Internet censorship and surveillance, and the International Telecommunications Union (ITU), which ranks countries based on their citizens' access and ability to use information and communications technology.[12]

In its 2011 survey, the OpenNet Initiative named thirty-three countries that have selective, substantial, or pervasive monitoring of political communications. Those countries range from very wealthy countries like Qatar, with an average per-capita GDP of $104,000, to poor ones like Yemen (per-capita GDP of $2,300).[13] As a group, those countries are poorer than the countries that do not restrict their citizens' communications. The thirty-three countries that ONI says are surveilling their citizens had an average per-capita GDP of $13,815. That's slightly above the global average, but still dramatically less than the countries that have free and open access to telecommunications.

As can be seen in the adjacent graphic, the countries with good access to telecommunications, as ranked by the ITU in 2011, are far richer than the global average. The top twenty-five countries on the ITU's list have an average per-capita GDP of $41,068, while the bottom twenty-five countries have an average per-capita GDP of just $1,616.[14] Furthermore, the countries that restrict access to the Internet (as tracked by ONI) or censor journalists (as tracked by the CPJ) are poorer than the countries where bits are allowed to flow Faster. The punch line here is apparent: free-flowing information is critical to human development and the creation of wealth.

People everywhere want to communicate. Just look at the incredible surge in mobile phone use; that surge is a direct result of Smaller Faster Cheaper phones. In 2000, fewer than one billion people on the planet had access to cell phones. By mid-2012, according to the World Bank, the world had more than six billion mobile phone subscribers. And of that number, nearly five billion were living in developing countries.[15]

The Faster the Bits the Wealthier the People

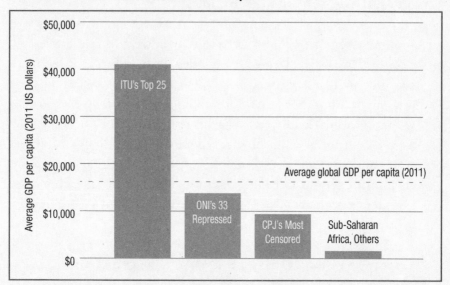

This graphic shows the correlation between economic well-being and information flows. The bar on the left shows the average per-capita GDP of the countries that rank highest in the International Telecommunications Union's survey of access to high-quality communication services. The bars to the right show the drastically lower incomes for people living in countries where information flows are restricted and journalists are censored. *Sources*: Country rankings from the International Telecommunications Union's ICT Development Index; ONI Political Monitoring of the Internet Index; and CPJ's Ten Most Censored Country Ranking.

Faster flows of information are also critical for education. As the speed of information flows on the Internet has increased, so have the educational opportunities. Broadband Internet access is allowing students all over the world to have access to libraries, articles, and instructors who just a few years ago were simply unavailable. The innovation that is occurring in online education is shaking up traditional education, and is doing so at a time when innovation in education is sorely needed.

Smaller Lighter Cheaper Phones

In 1984, New York City had about a thousand mobile telephones. Due to the lack of bandwidth, only a dozen of them could be used at any one time.[16] At that time, one of the only cell phone options was Motorola's DynaTAC 8000X, which was commonly known as "the brick." The phone weighed 2 pounds (907 grams), allowed about thirty minutes of talk time, and came with a price tag of $3,995.[17] (That's about $8,300 in current dollars.) In addition to its heft, the phone was huge, taking up nearly 80 cubic inches, or about 1,311 cubic centimeters.[18]

Today, one of the most popular mobile phones on the market is the Samsung Galaxy S3. It weighs 133 grams, allows 22 hours of talk time, and takes up about 83 cubic centimeters.[19] By late 2012, consumers could find unlocked Galaxy S3 phones on Amazon for about $600.[20] And some mobile phone companies were selling the phone for about $200 if the buyer agreed to a two-year contract.[21]

Thus, over the past three decades, an average mobile phone is about 16 times smaller and 7 times lighter than "the brick" of yesteryear. Better still, it's about 14 times cheaper and allows 44 times more talk time.[22]

15

FROM MONKS TO MOOCS

FASTER CHEAPER EDUCATION

In December 2012, Johnny Manziel, a quarterback from Texas A&M University, was awarded the Heisman Trophy, the most prestigious individual award in college football. He was the first freshman to win the Heisman. Two months later, Manziel announced that he would not be attending any on-campus classes at Texas A&M.

Manziel wasn't quitting school. Nor was he making a break for the National Football League. Instead, he was opting to do his coursework online. The move was made out of necessity. Manziel had enrolled in an on-campus English course at Texas A&M, but his celebrity disrupted the class. "It was a small class of 20 or 25—and it kind of turned into more of a big deal than I thought," he said.[1]

Precious few college students have to contend with the demands of autograph-seeking peers. But Manziel's decision to forgo on-campus classes in favor of online learning is part of trend that is shaking the education system to its roots. Thanks to the Internet, education is Faster Cheaper than ever before. Students interested in a certain subject, say, geometry, no longer need to reserve a spot in a class that meets at specific times at a specific location in order to study right angles, tangents, and parallelograms. With online education providers like Khan Academy, the geometry teacher—and all of the teacher's lectures—are available on demand, 24/7/365, in any location equipped with an Internet connection.

Online learning—whether it's a lesson on tying a bow tie or mastering differential equations—is giving more people access to high-quality education than ever before. We've gone from monks to MOOCs. We've gone from a system in which only the wealthy could afford tutors and books (dutifully hand-copied by monks) to a system where nearly everyone on the planet can, in theory, have access to some of the world's best teachers through massive open online courses (MOOCs). We've gone from the days of the aristocracy, where one educator taught one student, to MOOCs for the masses. In 2011, about 160,000 students in 190 countries enrolled in a free online course on artificial intelligence that was taught at Stanford University by two experts in the field, Peter Norvig and Sebastian Thrun.[2] That's not unusual. Another MOOC—also offered by Stanford in 2011, on machine learning—had a total class registration of 104,000; that's roughly twice the population of Enid, Oklahoma.[3] (For some of the leaders in online learning, see Appendix D.)

Of course, it's easier to register for a free class than it is to complete it. Of the 104,000 students who registered for Stanford's course on machine learning, only 13,000 completed the class work. Therein lies one of the many problems with MOOCs: high dropout rates. Add in the potential for cheating on exams, difficulty with proper grading, and the lack of personal interaction with the teacher, and the MOOCs begin to lose some of their shine. Perhaps the thorniest problem is accreditation. Sure, you may have completed the course on machine learning, but you didn't pay for it. And a school like Stanford, one of the most prestigious schools in the United States, is not eager to undermine its lofty reputation by allowing just anyone to claim that Stanford has stamped his or her educational passport. The move toward MOOCs may force schools to move away from a system based solely on traditional degree programs and toward a model based on competence. Some aspects of that system already exist with CLEP—short for College-Level Examination Program—which allows students to get college credit by passing tests that demonstrate their proficiency in various subjects.[4]

While there are many problems that must be addressed, it's also obvious that MOOCs and online learning are coming of age at the same time that much of the American education system—in particular, America's colleges and universities—is foundering on an outmoded business model, soaring costs, and woefully deficient results.

Student loan debt in the United States has surpassed $1 trillion, an amount that exceeds total US credit card debt. Tuition at both public and private schools continues to soar. A 2012 study by Bain & Company found that "annual tuition increases several times the rate of inflation have become commonplace" and yet, despite the higher costs being imposed on students, "a growing percentage of our colleges and universities are in real financial trouble."[5]

In 1970, the average tuition at a public, four-year college was $358 per semester and took about 4 percent of median family income. By 2010, that tuition figure had soared to $6,695, and the share of median family income had jumped to about 11 percent. If tuition increases had matched the rate of inflation over that same time frame, tuition would have been just $2,052.[6] (The rate of increase for private institutions was even higher than that for public schools.)

Furthermore, the size of the college-age population is falling.[7] In 2011, the number of US high school graduates peaked at 3.4 million and has begun to decline. That decline has resulted in declining college enrollments: in the first half of 2013, US college enrollment dropped by 2.3 percent compared with the same period in 2012.[7]

The emergence of online learning has led to some dire predictions. In 2013, Clayton Christensen, a professor at the Harvard Business School who has written extensively on his theory of "disruptive innovation," predicted that in "fifteen years maybe half of the universities will be in bankruptcy, including the state schools." Christensen said that for "three hundred years education wasn't disruptible because there wasn't a technological core." That is, there wasn't a new method of providing education services that could replace the old business model. Online learning, he said, brings a new technological core to the table and it "truly is going to kill us," with the "us" being traditional universities.[8]

Jeff Sandefer is an Austin-based entrepreneur and founder of two schools, the Acton School of Business and Acton Academy. The latter, which is an elementary and middle school, relies on Khan Academy in the classroom. Sandefer predicts that within a few years, "every elementary-school math teacher in America will be out of a job." While Sandefer may be overstating the case, the disruptive innovation now under way in the education business is unprecedented. Online education and MOOCs are catalyzing what may be the most important structural change in education since Jean-Baptiste de La Salle, a French cleric, began teaching teachers how to teach back in the seventeenth century.[9] De La Salle's heir apparent may be Salman Khan, a former hedge fund manager with two undergraduate degrees—in mathematics and electrical engineering—as well as a master's in computer science, all from MIT. He also has an MBA from Harvard.[10]

In 2004, Khan, who was born in New Orleans in 1976 to a Bangladeshi father and an Indian mother, began tutoring his cousin in math by using an online sketchpad provided by Yahoo! His tutorials proved popular with other friends and relatives, so Khan began putting his lessons on YouTube.[11] In 2006, Khan created Khan Academy, which is now supported by a variety of philanthropists, including the Bill & Melinda Gates Foundation.[12] Today, Khan's online school has more than 4,000 videos available on subjects ranging from geometry to art history. And the nonprofit isn't shy about its intent. "We're a not-for-profit with the goal of changing education for the better by providing a free world-class education for anyone anywhere."[13] That's a stunning vision, one that clearly depends on its students having access to a computer with broadband connectivity and a reliable flow of electricity.

Khan believes that the use of video on the Internet can reinvent education, and that reinvention appears to be well under way. By the end of 2012, the school's videos had been viewed more than 200 million times. Khan Academy is aiming to educate students all over the globe, with videos available in twenty-one different languages.[14]

The lower costs and ready availability of online courses have resulted in astounding growth. By 2013, more than five million students from around the world had registered for online classes.[15] And while the early results are promising, plenty of questions remain about online learning. Chief among them: the issue of human interaction. What works for teaching math or science online may not be effective when the subject is literature or drama. The give-and-take classroom discussions that have long been lauded as essential to the Socratic method cannot yet be duplicated online. As one analyst put it, the instructors who teach online courses are "simultaneously the most and least accessible teachers in history."[16] For example, students who sign up for class work with Coursera have been warned not to e-mail the professor or attempt to "friend" the teacher on Facebook.

The business model for online education must also evolve. Khan Academy may be able to survive by relying on philanthropy. But other providers are going to have to find ways to generate revenue from students who are used to getting nearly all of their information from the World Wide Web at no charge. Some online education providers are trying to address that problem by charging a fee to take the final exam for certain courses. By paying the fee and passing the exam, the students are then awarded a certificate confirming that they have completed the coursework.

Students still need mentors and the challenge of the give-and-take conversation that can be fostered in the classroom by a skilled teacher. The future of education, as David Brooks wrote in a 2013 column for the *New York Times*, may lie in differentiating between technical and practical knowledge.[17] Online courses are good at disseminating technical knowledge, such as how to solve quadratic equations or a given engineering problem. Face-to-face instruction is needed for learning more practical skills, such as how to negotiate a contract or argue a legal case.

MOOCs and online systems like Khan Academy are not a one-size-fits-all solution for education. But they are yet another indicator of how the Internet is reshaping our society and institutions. Clearly, online learning isn't limited to Khan Academy or courses on integral calculus. Thanks to YouTube, interested learners can get free online schooling in a raft of subjects, ranging from how to fix a bicycle tire to how to play the Hendrix Chord (E7 sharp 9) on the guitar.

Our interconnectedness is fostering new methods of teaching and learning that are Faster Cheaper than ever before. And many of the same innovations that are reshaping education—Cheaper computing and Faster broadband connectivity—are also reshaping health care.

16

SMALLER FASTER CHEAPER MEDICINE

Dr. Leonard "Bones" McCoy, the fictional doctor on the *Star Trek* TV series, would be impressed with the Scanadu Scout.

To be sure, the device introduced by California-based Scanadu in 2013 isn't as sophisticated as the fictional "tricorder" that McCoy used on *Star Trek* to instantly diagnose the maladies of his patients, but it nevertheless embodies the dramatic changes that are under way in medicine, a sector that is continually pushing for Smaller Faster Cheaper methods of doing business.

The Scout is about half the size of an iPhone. After being held on a patient's forehead for a few seconds, it provides the patient's body temperature, heart rate, blood oxygen level, respiratory rate, blood pressure, stress level, and even the electrical activity of the patient's heart. (Tracking that activity is known as electrocardiography.)[1] All of that data is then transmitted wirelessly from the Scout to the user's smart phone.

The Scanadu Scout obviates the need for numerous other devices. There's no need for a stethoscope to measure the patient's heart rate and respiratory rate. We can get rid of the sphygmomanometer to measure blood pressure. There's no need for a watch, thermometer, or electrocardiograph. Getting vital signs—a procedure that used to require a scheduled visit with a trained medical professional, who then needed several minutes to collect and record the data—can now be done with the Scanadu Scout almost anywhere, by anyone, in mere seconds. That data can then be shared with a nurse, physician, or maybe even a grandmother who's worried about the health of her daughter's baby.

Every day, in nearly every hospital in every country on the planet, medical professionals spend time collecting the vital signs of their patients. Thus, it's easy to imagine the amount of time that could be saved with the Scanadu Scout. Because the Scout records the data digitally, it allows practitioners to keep more accurate records and to manage that data Faster than ever before. And, of course, Faster nearly always means Cheaper.

The Scout is just one of an abundance of new medical technologies that are transforming an industry. Smaller cameras and surgical devices have led to a surge in the use of less-invasive techniques like arthroscopy and endoscopy. Smaller sensors and Cheaper electronics are making some medical devices wearable. Some are even edible. Add those improvements to the dozens of breakthroughs that have occurred in medicine over the past century or so, a list that includes antibiotics, X-rays, magnetic resonance imaging (MRI), and computed axial tomography (CAT) scans, and the progress becomes clear.

In his 2012 book, *The Creative Destruction of Medicine: How the Digital Revolution Will Create Better Health Care*, Dr. Eric Topol, a cardiologist at the Scripps Clinic and a professor of genomics at the Scripps Research Institute, declares that "medicine is about to go through its biggest shakeup in history." That shakeup, he says is due to "an unprecedented super-convergence" of digital technologies, including the "ubiquity of smart phones, bandwidth, pervasive connectivity, and social networking." Topol's book contains a graphic that excellently illustrates his point.

While Topol's graphic contains most of the factors that are driving changes in medicine, he forgot one of the most remarkable developments of the Internet Age: the ability of entrepreneurs to raise money from people they don't even know. And that takes us back to Scanadu. In May 2013, the company announced a campaign to raise $100,000 on the site Indiegogo to help fund development of the Scout. (Indiegogo is one of several crowd-funding operations. Kickstarter is another.) Within a few hours of posting its project on Indiegogo, Scanadu had met its $100,000 goal. By October 2013, it had raised $1.6 million.[2]

While it's certainly true that the American health-care system is, in many cases, absurdly expensive and in desperate need of reform,

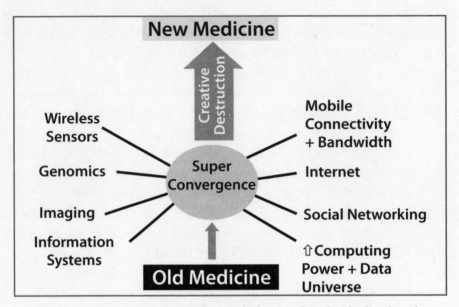

Physician/futurist Eric Topol foresees a future of "individualized medicine that is enabled by digitizing humans." *Source: The Creative Destruction of Medicine* by Eric Topol. Reproduced with permission.

it's also true that many procedures have gotten dramatically Cheaper. For instance, between 1910 and 2012, the cost of an average X-ray has declined by about 80 percent.[3] And unlike the clunky X-ray machines of yesteryear, patients and doctors can now use full-motion X-ray machines. Thanks to a Florida-based company called Digital Motion X-Ray, doctors can look at a patient's hand, knee, spine, or other body part, and examine it as it moves. This technology gives doctors a better understanding of the area being examined.[4]

The first sequencing of the human genome was completed in 2003. It took nearly a decade and cost about $3 billion to accomplish that feat.[5] But the costs to do the sequencing are falling dramatically. By 2012, the cost of sequencing a human genome had dropped to about $5,000.[6] A commonly stated goal is to be able to do it for $1,000. The ability to have low-cost DNA information about a patient could allow doctors to quickly customize treatment plans based on the patient's genetic makeup. And the time needed to sequence DNA is declining. A company called Ion Torrent is using semiconductors and ion spectroscopy so that it can now decode DNA in just two hours.[7]

Speaking of Cheaper, consider 23andMe. For $99, 23andMe will send you their "DNA spit kit," which you can use to provide the company with a saliva sample. In four to six weeks, the company will send you a report on your genetic makeup, which includes details on your ancestry as well as info on your predisposition to a variety of possible health risks such as heart disease and breast cancer.[*]

Smaller Cheaper sensors could help us in the fight against diabetes. About nineteen million Americans have been diagnosed with the disease, and another seven million or so have diabetes but haven't yet been diagnosed.[8] For diabetics, maintaining the proper level of blood glucose can mean the difference between life and death. For years, the only way to monitor blood glucose was by pricking a finger and doing a blood assay, a messy and time-consuming process. But a new technology that relies on a tiny sensor implanted below the skin of the abdomen could make that messy procedure a thing of the past. The technology—continuous glucose monitoring—allows diabetics to use a small electronic device that wirelessly receives data from the implanted sensor. The device provides a readout of blood glucose levels and sounds an alarm if levels get too high or too low.[9]

Commonly used medical technologies now include an endoscopic capsule, a device equipped with its own light, battery, and camera, which transmits video images from inside the body. After being swallowed by a patient, the device can provide doctors with real-time images from inside the patient's digestive tract.[10] (As an example of how hot this sector is: in late 2013, Dublin-based health-care equipment company Covidien announced it would buy Given Imaging Ltd. for $860 million. The deal gives Covidien control over Given Imaging's PillCam

---

[*] In November 2013, the Food and Drug Administration ordered 23andMe to stop selling its DNA test kits because the agency claimed the company did not have "marketing clearance." For more, see: http://www.fda.gov/ICECI/EnforcementActions/Warning Letters/2013/ucm376296.htm.

capsule endoscopy technology, which has been used more than two million times.)[11] Or consider how dentists and surgeons are using scanners and 3-D printers. Dentists are already using those tools to quickly produce dental crowns. Surgeons are testing 3-D printers with the idea that they may be able to custom-print joints—knees are an early focus of their work—for their patients.[12]

The push to develop Faster Cheaper medical technologies comes not only from the desire to provide better patient care, but also to save money. In 2012, the United States spent more than $8,200 per capita on health care. That figure is more than twice as much as is spent by Western European countries such as France, Sweden, and the United Kingdom. For comparison, Australia spends less than $3,700 per person per year on health care. Nearly 18 percent of the US GDP—about $2.7 trillion per year—is now spent on health care.[13] Add in the trillions of dollars being spent by other countries, and the size of the market becomes even more obvious. Furthermore, demographic trends point toward an aging population that will require more medical care in the years ahead.

Given those trends, the revolution that Topol envisions in *The Creative Destruction of Medicine* is dearly needed. Such a revolution won't just be good for patients; it will also benefit medical professionals and the broader economy.

While the need for more effective tools like the Scanadu Scout and better medicines is obvious, we often overlook the fact that many of our most important medical devices and institutions rely on cheap abundant energy. Hospitals must be able to operate 24/7. And to provide the best medical care, they need super-reliable flows of electricity.

The next section discusses the urgent need for Cheaper energy.

PART III:

The Need for Cheaper Energy

17

THE FASTER THE (DRILL) BITS, THE CHEAPER THE ENERGY

When asked what he liked most about the new drill rig he was working on, Artie White had a quick reply: "They are so much safer than the old ones."

White has spent much of his adult life working on drill rigs. As a roughneck, he worked countless hours on the main floor of drilling rigs in all kinds of weather—heat, cold, rain, snow. Heavy equipment was always overhead. The spinning drill pipe and numerous heavy tools on the deck meant there was near-constant danger of getting pinched or crushed. Despite constant attention, the drilling floor was nearly always slick. Drilling fluid that splattered off the drill pipe, as well as condensation and occasional rain, required workers on the rig to be continually aware of their footing.

But on a sunny, windy day in January 2013, White wasn't swinging tools on the drill rig floor. There wasn't a speck of dirt on his gray overalls. He wasn't wearing a hard hat or safety glasses. Instead, White was drilling a well known as the Tom Horn 7–13–9 10H—located about 5 miles north of the old Route 66 in Canadian County, Oklahoma—while sitting in a climate-controlled booth in a comfortable chair. The well was one of dozens in the region that Oklahoma City–based Devon Energy was drilling in the Cana Woodford Shale, a huge formation of sedimentary rock about 13,000 feet below where White and I were standing.

No longer a roughneck, White, a native of Purcell, Oklahoma, had advanced to the position of driller. He was the foreman on the rig and was responsible for all of the activities on the drilling floor.[1] Four

January 26, 2013: Artie White, the driller on Helmerich & Payne's Flex-Rig 268, stands at his workstation while drilling a well known as the Tom Horn 7–13–9 10H. The outline of the drill pipe and drive mechanism can be seen behind him. The AC top-drive rig represents a major step forward in the safety and digitization of the drilling process. *Source:* Photo by author.

workers, including two roughnecks, reported to him. To his right was a joystick similar to what you'd find in a video arcade. To his left, a bank of flat-screen monitors provided him with the rig's critical information: the depth of the bit, the speed at which it was rotating, the weight of the drill pipe in the well, the flow rate of the drilling fluid (known as mud) that was being pumped into the bottom of the hole, and other data.

Controlling most of the drill rig's critical functions with digital controllers means there are "a lot less ways to get hurt," said White, who, as he talked, kept returning his eyes to the rotating drill pipe, which was about 15 feet north of him and could easily be seen through the thick glass window behind his workstation. It's no surprise that safety is a prime concern for White. The people who work on drilling rigs have long had some of the hardest and most dangerous jobs in America.[2] The rig that White and his coworkers were using—Tulsa-based Helmerich & Payne's FlexRig number 268—had all of the latest equipment, including a hulking apparatus on the deck of the rig called an "iron roughneck," a robotic device that performs many of the more mundane—and dangerous—procedures that used to be done by humans.[3]

While safer working conditions are critical to every industry, the safety improvements on drill rigs are only part of the story. FlexRig number 268 is also a sixteen-story-tall example of the never-ending push for Smaller Faster Lighter Denser Cheaper in the Oil Patch. Silicon and software have replaced human interventions for the rig's key functions. That allows the rig to drill more wells, Faster. Given that operating a land rig may cost $4,000 per hour, Faster drilling means Cheaper drilling.[4]

Over the past century, oil and gas drilling has gone from a business dominated by wildcatters armed with a hunch and a prayer to one that is more akin to the precision manufacturing that dominates aerospace and automobiles. Today, drillers like Artie White are so good at what they do that they can punch holes in the earth that are 2 miles deep, turn their drill bit 90 degrees, drill another 2 miles horizontally, and arrive within a few inches of their target.

In January 2011, during his State of the Union speech, President Barack Obama called oil "yesterday's energy."[5] That sound bite may appeal to the noisy members of the Green Left who are advocating for more mandates and subsidies for solar and wind energy, but here's the reality: oil has been "yesterday's energy" for more than a century. And yet, it persists. It persists because of continuing innovation that allows drillers like Artie White to produce more oil and gas and do so Faster Cheaper.

If oil didn't exist, we would have to invent it. No other substance comes close to oil when it comes to energy density, ease of handling, and flexibility. Those properties explain why oil provides more energy to the global economy than any other fuel.[6] (Oil provides about 33 percent, coal provides 30 percent, natural gas provides 24 percent, and hydro, 7 percent. The balance comes from nuclear and renewables.) It also explains why more than 90 percent of all transportation continues to be fueled by petroleum products.

Oil—and natural gas—are going to continue dominating the global energy market for decades to come because companies like Helmerich & Payne, Devon Energy, and dozens of others have a simple choice: innovate or die. Innovate they have. The convergence of several technologies ranging from better drill bits and seismic techniques to robotic rigs and nanotechnology are allowing the oil and gas sector to produce ever-increasing quantities of energy at lower cost. Furthermore, those technical advances are being deployed by an industry that is spending enormous sums every year to find, refine, and transport the fuel that the world's consumers demand.

Advocates of solar, wind, and other renewable technologies like to point to the rapid growth that has occurred in the solar- and wind-energy sectors over the past few years. In 2011, Bloomberg New Energy Finance estimated that global investment in "clean energy" totaled $302 billion, a record. (In 2012, investment fell slightly to $268.7 billion.)[7] That's a lot of money. But in 2012, global spending on oil and gas drilling totaled more than $1.2 trillion.[8] About a quarter of that amount, roughly $300 billion per year, is being spent drilling wells in the United States. Thus, every year, America alone is spending as much just drilling oil and gas wells as the entire rest of the world is spending on "clean energy." And drillers are getting more efficient all the time, as shown, once again, by the numbers.

More wells have been drilled in the United States than anywhere else. No other country even comes close. Between 1949 and 2010, more than 2.6 million oil and gas wells were drilled in the United States, and that number has been increasing by about 41,000 new wells per year. And yet, thanks to ongoing innovation in the oil field, between 1949 and 2011, the percentage of dry wells, or "dusters" dropped from 34 percent to 11 percent.[9]

Perhaps the most important recent innovation in the Oil Patch has occurred in the design of drilling rigs. Drill rigs are relatively simple devices. They can range in size from small truck-mounted units that

**Number of US Oil and Gas Wells Drilled and Percentage of Dusters
1949–2010**

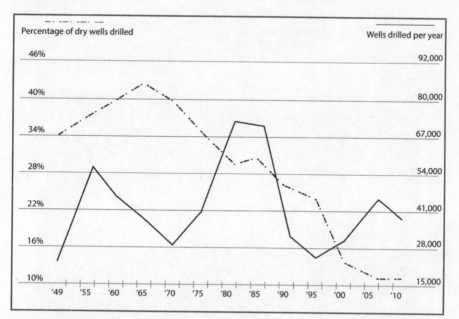

Source: Energy Information Administration (EIA).[10]

drill water wells to massive 100,000-ton drill ships that can prospect for
oil and gas in 10,000 feet of water.[11] But the principles are basically the
same: they must be stable, precise, and capable of producing the torque
needed to punch a hole in the earth. The latest, most important innovation in drill rigs is the AC top-drive. That's the design that Artie White
and his colleagues were using to drill the Tom Horn 7–13–9 10H. The
key improvement came by moving the rig's main drive mechanism
from the floor of the rig onto the mast. The result is a major step forward in the speed and control of the drilling process.

The AC top-drive consolidates the rig's drive and hoist mechanism
into one unit. That allows the automation of several mundane processes that used to require human intervention. Although many of the
operations on the rig still must be handled by roughnecks, a bank of
computers monitor key data points such as rotational speed on the bit,

January 26, 2013: A pair of roughnecks on the floor of an AC top-drive rig in Canadian County, Oklahoma. The lower part of the 800-horsepower drive mechanism can be seen above their heads. The rig is powered by three diesel engines that produce a total of 4,428 horsepower. Those engines produce the electricity needed to run nearly everything on the rig: the motors, lights, computers, hoist, and drive mechanisms. *Source:* Photo by author.

drilling rate, and flow rates. The computers feed that data into a drilling control system that keeps the optimum amount of weight on the drill bit and keeps it spinning at optimal speed.

If you've ever drilled a hole in Sheetrock or a piece of wood, you know that applying proper pressure is key. Press too hard, and the drill freezes or gets stuck. Not enough pressure or insufficient speed, and the drill bit makes no progress. The same factors are at play on a drill rig that's boring a four-mile-long well. By moving the rig's prime mover from the floor to the mast, the AC top-drive allows digital controllers to continually weigh the entire drill string. By monitoring the weight, the system continually adjusts the amount of pressure being applied to the bit, as well as the rotational speed. Those adjustments assure the maximum rate of penetration. Add in the rig's ability to use longer sections of drill pipe and its modular design—which allows it to be transported more quickly than older rig designs—and it's easy to see how companies are able to drill more wells Faster Cheaper.

In addition to better drill rigs, we've seen innovation in drill bits, seismic tools, telemetry systems, proppants, pumps, and numerous other technologies that are needed to produce hydrocarbons. All of those innovations have resulted in big improvements in speed. Back in 2007, Devon Energy needed about fifty-seven days to drill an average well in the Cana Woodford Shale. By 2012, the company was able to drill a well like the Tom Horn 7–13–9 10H—a well with a vertical depth of 12,814 feet and a lateral extension of nearly 5,000 feet—in just thirty days.[12]

Other drillers are showing similar speed improvements. Southwestern Energy is a Houston-based company that has pioneered the development of the Fayetteville Shale in Arkansas. Between 2007 and 2012, the cost of an average well that Southwestern drills in the Fayetteville Shale has stayed fairly constant, at about $3 million per well. But over that same time frame, Southwestern reduced the number of days needed to drill a well from seventeen days to just seven days. Better yet, the initial production rate on the wells being drilled has more than tripled.[13]

The result of all that innovation can be seen in the production data. In 2012, US oil production rose by 790,000 barrels per day, the biggest annual increase since US oil production began in 1859.[14] Domestic

On January 26, 2013, a drill bit made by Halliburton sits in the foreground of the well known as the Tom Horn 7–13–9 10H. Polycrystalline diamond compact bits like this one are often used in place of traditional roller-cone drill bits when drilling in shale formations. Renting a PDC often costs about $20,000. Drilling the Tom Horn 7–13–9 10H required the use of about ten bits like this one. *Source:* Photo by author.

natural gas production is also at record levels. The United States has been the epicenter of oil and gas exploration and production for more than a century. And yet, thanks to ongoing innovation, production keeps rising.

The extension of the hydrocarbon era is happening thanks to the industry's innovation onshore and offshore. One of the biggest offshore discoveries in recent years was the Johan Sverdrup field in the North Sea. The Sverdrup field alone contains up to 3.3 billion barrels of recoverable hydrocarbons, making it the largest discovery in the North Sea since 1980.[15] In early 2013, a pair of offshore discoveries in the Lower Tertiary Trend in the US Gulf of Mexico confirmed the presence of billions of additional barrels of oil and natural gas resources.[16] Few countries provide a better demonstration of offshore oil innovation

Offshore Oil and Gas Discoveries, 1995–2012

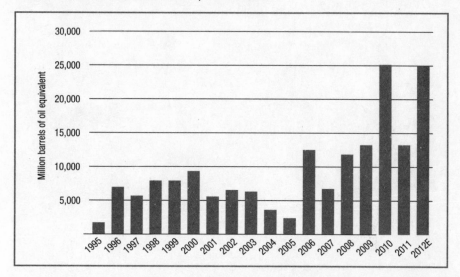

The innovations that are happening in onshore drilling are being mir-
rored by continuing discoveries offshore. Advances in materials science,
robots, submarines, and other technologies are allowing companies to
access offshore oil and gas reservoirs that were once thought inaccessible.
As this graphic shows, between 2002 and 2012, more than 100 billion
barrels of new oil resources were discovered in offshore locations around
the world. In 2012 alone, global offshore oil discoveries totaled some 25
billion barrels. *Source*: Deutsche Bank and Wood MacKenzie.[17]

than Brazil. In 1990, Brazil, the largest country in South America, was
producing 650,000 barrels of oil per day. In 2011, production had increased to nearly 2.2 million barrels per day. Nearly all of that increased
production came from offshore wells.[18]

In 1929, the economic historian Abbott Payson Usher wrote: "The
limitations of resources are relative to the position of our knowledge
and of our technique." He continued, explaining that the perceived
limits of available resources "recede as we advance, at rates that are
proportionate to the advance in our knowledge."[19] The history of the
oil and gas sector is one of advancing knowledge and increased resource
availability. And that has resulted in Cheaper energy.

We're Running Out of Oil . . .

In 1914, a US government agency, the Bureau of Mines, predicted that world oil supplies would be depleted within ten years.

In 1939, the US Department of the Interior looked at the world's oil reserves and predicted that global oil supplies would be fully depleted in thirteen years.[20]

In 1946, the US State Department predicted that America would be facing an oil shortage in 20 years and that it would have no choice but to rely on increased oil imports from the Middle East.[21]

In 1951, the Interior Department said that global oil resources would be depleted within thirteen years.[22]

In 1972, the Club of Rome published *The Limits to Growth*, which predicted that the world would be out of oil by 1992 and out of natural gas by 1993.[23]

In 1974, population scientist Paul Ehrlich and his wife, Anne, predicted that "within the next quarter of a century mankind will be looking elsewhere than in oil wells for its main source of energy."[24]

Reality check: in 1980, the world had about 683 billion barrels of proved reserves. Between 1980 and 2011, residents of the planet consumed about 800 billion barrels of oil. Yet in 2011, global proved oil reserves stood at 1.6 trillion barrels, an increase of 130 percent over the level recorded in 1980.[25]

We're Also Running Out of Natural Gas . . .

In 1922, the US Coal Commission, an entity created by President Warren Harding, warned that "the output of [natural] gas has begun to wane."[26]

In 1956, M. King Hubbert, a Shell geophysicist who became famous for his forecast known as Hubbert's Peak, predicted that gas production in the United States would peak at about 38 billion cubic feet per day in 1970.[27]

In 1977, John O'Leary, the administrator of the Federal Energy Administration, told Congress that "it must be assumed that domestic natural gas supplies will continue to decline" and that the United States should "convert to other fuels just as rapidly as we can."[28]

In 2003, Matthew Simmons, a Houston-based investment banker for the energy industry who was among the leaders of the peak oil crowd, predicted that natural gas supplies were about to fall off a "cliff."[29]

In 2005, Lee Raymond, the famously combative former CEO of ExxonMobil, declared that "gas production has peaked in North America."[30]

Reality check: in 2012, US natural gas production averaged 69 billion cubic feet per day, a record, and a 33 percent increase over the levels achieved in 2005, when Raymond claimed North American production had peaked.[31]

18

THE TYRANNY
OF DENSITY

Among the Mount Everest of inanities ever uttered on the subject of energy, the blue-ribbon winner must be this one: "the tyranny of oil."

Both Barack Obama and Robert F. Kennedy Jr. have used the line. Obama claimed it for his own in 2007 during a speech in which he declared his run for the White House. While standing on the steps of the Old State Capitol in Springfield, Illinois, Obama said, "Let's be the generation that finally frees America from the tyranny of oil."[1]

In March 2013, during a speech at Sandhills Community College in North Carolina, Kennedy, a high-profile opponent of the Keystone XL pipeline (he was arrested at the White House during an anti-Keystone protest), said, "we need to free ourselves from the tyranny of oil."[2]

That Obama and Kennedy, both of whom went to Harvard, claim that a super-high-energy density substance that can be deployed for innumerable purposes, from pumping well water in Kenya to emergency generation of electricity in Lower Manhattan, is somehow bad or even yet, tyrannical, is nonsense on stilts. Rather than talk about the tyranny of oil, the two Harvard grads might as well complain about the tyranny of physics—or better yet, the tyranny of density.

Few substances this side of uranium come close to touching oil when it comes to the essential measure of energy density: the amount of energy (measured in joules or BTUs) that can be contained in a given volume or mass. In addition to petroleum's high energy density, it is stable at standard temperature and pressure, relatively cheap, easily transported, and can be used for everything from making shoelaces to fueling jumbo jets.

Oil's tyranny of density can be demonstrated by looking at the aviation sector and by doing a tiny bit of math. To make it easy, let's use metric units and focus on weight, as that factor is critical in aerospace. The gravimetric energy density of jet fuel is high: about 43 megajoules (million joules) per kilogram. (Low-enriched uranium, by the way, is 3.9 terajoules—trillion joules—per kilogram.)[3]

Keep those numbers in mind as we look at the best-selling jet airliner in aviation history: the Boeing 737.[4] A fully fueled 737–700 holds about 26,000 liters of jet fuel, weighing about 20,500 kilograms. That amount of fuel contains about 880 gigajoules (billion joules) of energy. The maximum take-off weight for the 737–700 is about 78,000 kilograms; therefore, jet fuel may account for as much as 26 percent of the plane's weight as it leaves the runway.[5]

Obama and Kennedy are big fans of electric cars.[6] Lithium-ion batteries have higher energy density than most other batteries, holding about 150 watt-hours—540,000 joules—of energy per kilogram.[7] Recall that jet fuel contains about 43 megajoules per kilogram, or nearly 80 times as much energy. Therefore, if Boeing were trying to replace jet fuel with batteries in the 737–700, it would need about 1.6 million kilograms of lithium-ion batteries. That means that the 737–700 would require a battery pack that weighs about 21 times as much as the airplane itself.

Prefer to use a "green" fuel like firewood? With an energy density of about 16 megajoules per kilogram, that same 737–700 would require about 55,000 kilograms of wood. With that much kindling onboard, rest assured there won't be room in the overhead bin for your carry-on bag.

Even at 35,000 feet, the truth is obvious: the only tyranny at work in our energy and power systems is that of simple math and elementary-school physics. Obama and Kennedy may not like oil, and their allies on the Left may hate Shell/BP/Marathon/Exxon/Saudi Aramco/Chevron/Keystone XL, but here's the reality: oil is a miracle substance. Without it, modern society simply would not be possible.

Rather than condemning the fuel that makes modern life possible, our political leaders should be figuring out how we can make oil more available to more people at lower cost.

Denser Energy Is Green Energy:
Comparing Uranium with Various Other Sources

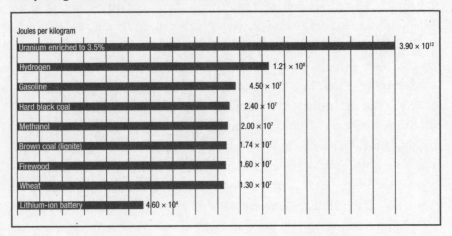

Joules per kilogram

Uranium enriched to 3.5%	3.90×10^{12}
Hydrogen	1.21×10^{8}
Gasoline	4.50×10^{7}
Hard black coal	2.40×10^{7}
Methanol	2.00×10^{7}
Brown coal (lignite)	1.74×10^{7}
Firewood	1.60×10^{7}
Wheat	1.30×10^{7}
Lithium-ion battery	4.60×10^{4}

We use oil because of its high-energy density and other characteristics. But gasoline—and even hydrogen—are no match for uranium, which when enriched to 3.5 percent, has 3.9 terajoules per kilogram. For comparison, firewood has about 16 megajoules per kilogram. Thus, to match the energy of one kilo of enriched uranium, you'd need about 244,000 kilos of firewood. *Source:* World Nuclear Association.[8]

CLEAN ENERGY SYSTEMS

Official Name Clean Energy Systems, Inc.

Web site http://www.cleanenergysystems.com

Ownership Private

Headquarters Rancho Cordova, California (US)

Amount of Capital Raised $100 million

Latest Annual Revenue N/A

Clean Energy Systems is bringing rocket technology to the electricity business. In doing so, it is achieving power densities that approach those seen in nuclear reactors.

Clean Energy Systems is proving its Smaller Denser technology amid the dusty almond groves that lie on the outskirts of Bakersfield, California. On the site of a former biomass-to-energy plant, the company is testing a gleaming prototype called the 12-inch Gas Generator, which is, in effect, a modified rocket engine that has been laid horizontally. The machine is a marvel of precision engineering, curved pipes, stout bolts, and burly flanges. On a hot day in mid-2013, the company's plant manager, Heath Evenson, explained the process. Natural gas and pure oxygen are ignited under high pressure at the top of the engine. Just downstream of where the ignition occurs, a series of injection nozzles spray water into the combustion chamber. The exhaust gas—a mixture of nearly pure carbon dioxide and high-pressure

June 27, 2013: Heath Evenson, the plant manager at Clean Energy Systems' plant in Bakersfield, California, shows visitors the company's newest design: a machine that uses rocket technology to produce high-pressure, high-temperature steam that, in turn, can be used to generate electricity. *Source:* Photo by author.

steam—then flows into a generation unit that produces electricity. By adjusting the flow of fuel, oxygen, and water, Clean Energy Systems can change the outlet pressure as well as the temperature of the steam being produced. That flexibility means that the company's generation units could be employed for a variety of uses, including electricity generation, refining, and enhanced oil recovery.

What's intriguing about the design is its compactness. The essential components are small enough to fit inside a single 40-by-8-foot shipping container. Under optimum conditions, the machine can generate about 70 megawatts of electricity. When accounting for all of the equipment needed by the system, including the air separators needed to produce oxygen, Clean Energy

Systems' generator has an areal power density of about 117,000 watts per square meter.[9] That's an astoundingly high number, particularly when you consider that wind energy has a power density of 1 watt per square meter, and even the best solar systems are in the low double digits.

Clean Energy Systems expects to begin commercial deployment of its first units in 2014 or 2015. The first applications are likely to be for use in oil-recovery systems that need steam and/or carbon dioxide. Among the likely locations: the Permian Basin in Texas, a region chronically short of high-quality carbon dioxide, a gas that is valuable for enhanced oil recovery.

19

SMALLER FASTER
AND THE
COAL QUESTION

About one train per hour: that's the target loading rate for the massive coal silos, conveyors, and hoppers at the North Antelope Rochelle Mine in central Wyoming's Powder River Basin.

And on a cool, nearly windless day in March 2012, Scott Durgin, a regional vice president for Peabody Energy, was happy. Standing in the mine's dispatch office, one wall of which was covered with flat-panel displays showing details of nearly every facet of the mine's operation, Durgin pointed to a list of trains that had recently passed through the mine. A clock on the wall displayed exactly 12 noon. The huge mine had loaded eleven trains—each one carrying about 16,000 tons of coal—in the previous twelve hours.

Some 40 percent of all US coal production comes from the strip mines located within a few dozen miles of the control room where we were standing. When I asked Durgin how long Peabody could continue mining in the region, he replied that the company could easily keep going for another five decades. "There's no end to the coal here," he said.

The scale and productivity of the mine is difficult to imagine. It produces about 3 tons of coal *per second*. The coal from the mine could, on an ongoing basis, provide enough coal to generate about two-thirds of Mexico's electricity needs.[1]

Despite its staggering productivity, the North Antelope Rochelle Mine—along with all of the other 1,300 coal mines operating in the United States—should be an anachronism.[2] Today's TV commercials

March 29, 2012: A Peabody Energy employee looks over the working face of the North Antelope Rochelle Mine. The coal seam being mined at the Wyoming site is about 80 feet thick. *Source:* Photo by author.

and talking heads lionize wind turbines and solar panels, not coal mines. Indeed, coal is routinely demonized. In 2009, climate scientist James Hansen wrote an opinion piece for the *Observer*, in which he declared that "coal is the single greatest threat to civilization and all life on our planet . . . Trains carrying coal to power plants are death trains. Coal-fired power plants are factories of death."[3]

Although the anti-coal forces are strong, coal will not go away, because coal is a fuel of necessity for producing electricity. Demand for electricity throughout the world is soaring. And as coal deposits are abundant, widely dispersed, easily mined, and are not influenced by any OPEC-like entities, it's easy to see why coal is the world's fastest-growing energy source. Between 2002 and 2012, when US coal consumption fell by about 21 percent, global coal consumption soared, growing by 26.5 million barrels of oil equivalent per day. That nearly matches the

Global Coal Consumption 1980–2011, and Projected to 2035

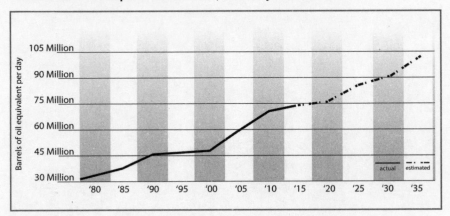

The ongoing boom in coal consumption in India, China, Germany, and other countries assures that coal use will continue rising for years to come. Between 2012 and 2035, the EIA expects coal consumption to rise by about 38 percent to about 98.5 million barrels of oil equivalent per day. *Source*: EIA.[4]

global growth in consumption of oil, natural gas, nuclear, and wind energy *combined* over that time.[5]

And therein lies a remarkable dichotomy: at a time when nearly everything in modern society is getting Smaller Faster Lighter, coal—a bulky, heavy fuel that's more identified with the nineteenth century than the twenty-first—remains a remarkably resilient source of energy. Coal may be bulky and heavy, but it's an excellent fuel for creating electrons, and few things are Smaller Faster Lighter than electrons. (Electricity travels at about the speed of light, which is roughly 1 billion kilometers—670 million miles—per hour.)

To be perfectly candid, the coal boom is problematic for the thesis of this book. The black fuel's persistence in our energy mix both beggars and confirms the thesis at hand. Oil and natural gas are vastly superior in terms of their hydrogen-to-carbon ratios. For instance, diesel fuel contains about 60 percent more energy per kilogram than the best black coal; natural gas has about 90 percent more.[6]

1906: Miners use hand tools while mining coal by candlelight. *Source*: Library of Congress, LC–USZ62–27886.

But coal—stubborn old coal, a fuel that's been used by humans for millennia, and now accounts for about 40 percent of all global electricity production—enables innovation.[7] Without cheap supplies of electricity produced from coal, the ongoing revolution in information technology, as well as the age of biotech and nanotech, simply wouldn't be possible. Electricity accelerates the trend toward objects and systems that are Smaller Faster Lighter Denser Cheaper.

Electricity is the fuel of modernity. Countries that have cheap, abundant, reliable supplies of electricity are able to bring their people out of darkness and poverty and into the light of the modern world. And from India and China to Vietnam and Indonesia, the fuel that's supplying the vast majority of that electricity is coal.

Coal has been an essential fuel for electricity production ever since Thomas Edison used it in the first central power plant in Lower

Manhattan in 1882. Of course, Edison wasn't the first to exploit coal. The first instance of coal use likely occurred in China about 3,000 years ago. By the late 1600s, the town of Birmingham, England, had become a center of the metal-working business. The town had more than two hundred forges that used coal to produce iron.[8]

For much of human history, the black fuel has engendered an intense love-hate relationship. Coal heated people's homes and fueled the Industrial Revolution in England, but it also made parts of the country, particularly the smog-ruined cities, nearly uninhabitable. In 1812, in London, a combination of coal smoke and fog became so dense that according to one report, "for the greater part of the day it was impossible to read or write at a window without artificial light. Persons in the streets could scarcely be seen in the forenoon at two yards distance."[9] Today, two hundred years later, some of the very same problems are plaguing China. In Datong, known as the "City of Coal," the air pollution on some winter days is so bad that "even during the daytime, people drive with their lights on."[10]

The air pollution that comes with today's coal business is only part of the industry's environmental and human toll. The coal industry causes serious damage on the surface of the Earth through strip mines, mountaintop removal, and ash ponds at power plants. In addition, thousands of miners die each year in the world's coal mines, with Chinese mines being the most deadly. Here in the United States, forty-eight coal miners were killed in 2010, with twenty-nine of those deaths due to an explosion at the Upper Big Branch mine in West Virginia.[11] That explosion was the worst mining accident in the United States in four decades.[12]

While air pollution from burning coal and the problems associated with coal production are deadly serious, the coal-related issue that's getting the most attention from the media and policy makers is carbon dioxide emissions and climate change. While the US EPA and the Obama administration may want to prohibit the construction of new coal-fired power plants, their efforts will have almost no effect on global carbon dioxide emissions. Why? The United States may quit burning coal, but the rest of the world can't, and won't.

India Is *Not* Going "Beyond Coal"

In July 2012, blackouts hit northern India, leaving more than 600 million people—about twice the population of the United States—without electricity. Trains were stranded, traffic snarled, and the country's economy ground to a halt. The blackouts were caused by excess demand when some regions of the country began taking more power than they had been allotted.[13] In the months since the blackout, one thing has become certain: India won't be going "beyond coal" anytime soon.

At the same time that the Sierra Club pushes its "beyond coal" campaign in the United States, developing countries are rapidly increasing their coal consumption. Much of that surge in coal use occurred in India, the world's third-largest coal consumer (behind only China and the United States). Burgeoning coal use helps explain why India's carbon dioxide emissions jumped by 81 percent between 2002 and 2012. That same coal use explains why global carbon dioxide emissions continue to soar.[14]

Although India's coal use—which doubled between 2002 and 2012 to some 6 million barrels of oil equivalent per day—keeps rising, the country remains chronically short of electricity. India's per-capita electricity consumption is about 700 kilowatt-hours per year.[15] For comparison, the average resident of China uses almost five times as much electricity as the average Indian, while the average American uses about 19 times as much.[16] In mid-2013, Victor Mallet of the *Financial Times* reported on India's electricity shortages: "Of all the problems blamed for the slowdown [of economic activity] over the past two years . . . the electricity shortage is now regarded by government and business alike as among the most serious." Mallet went on to quote a government official who said that the country's leaders used to think roads were the key to growth. But now, said the official, "it's power, power, power." To alleviate the shortages, India is planning to add about 90,000 megawatts of new generation capacity by 2018.[17]

While India wants to increase the amount of its natural gas–generation and nuclear-generation capacity, it still relies on coal for about two-thirds of its electricity production.[18] With 60 billion tons of domestic coal reserves—enough to last a century at current rates of extraction—India has plenty of the carbon-heavy fuel. But the country's mines are notoriously inefficient, and coal deliveries have been hamstrung by poor-quality transportation and ham-handed government policies. The result: India, which now imports about 25 percent of its coal, may soon surpass China as the world's biggest coal importer.[19]

For years, Indian leaders have been saying that they will not let concerns about carbon dioxide impede their push to generate more electricity. In 2009, shortly before the big climate-change meeting in Copenhagen, that very message was delivered by none other than Rajendra Pachauri, the Indian academic who chairs the UN's Intergovernmental Panel on Climate Change. "Can you imagine 400 million people who do not have a lightbulb in their homes?" he asked. "You cannot, in a democracy, ignore some of these realities and as it happens with the resources of coal that India has, we really don't have any choice but to use coal."[20]

Use it they will. In the wake of the blackouts, Indian officials are talking about expediting the permits needed to produce and transport more coal. Over the next decade or two, India's coal use is expected to double, and so will easily surpass US coal consumption.[21] But even if that occurs, India will likely continue lagging the developed world in production of electricity.

Slogans like "beyond coal" may appeal to members of the Sierra Club and to former New York Mayor Michael

Bloomberg, who gave the environmental group $50 million to help it "end the coal era."[22] But with 1.3 billion people on the planet still lacking access to electricity, the priority for leaders in New Delhi and other developing countries isn't carbon dioxide emissions or "clean energy." Instead, their primary aim is to simply bring their people out of the dark.

The rest of the world will keep burning coal because the essentiality of electricity to modernity is incontrovertible. The countries that have cheap, abundant, reliable electricity are wealthier than the countries that don't. And the people who live in wealthier countries are living better, longer, healthier lives than they ever have before. The International Energy Agency calls electricity "crucial to human development," and it says that the availability of electricity is "one of the most clear and un-distorted indications of a country's energy poverty status."[23]

Perhaps the best example of soaring electricity demand can be found in Vietnam. Between 2001 and 2010, electricity use in that country soared by 227 percent, a rate Faster than that of any other country on

Electricity Use Is Closely Correlated with Wealth Creation

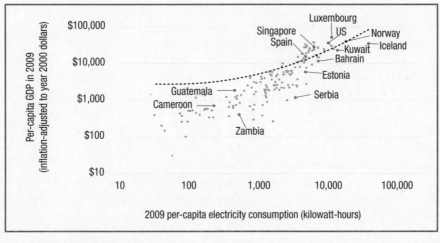

Source: World Bank.[24]

the planet.[25] The country's soaring electricity demand is also stoking demand for coal. Between 2001 and 2010, the country's coal demand increased by 175 percent, again, Faster than in any other country. That coal use also led Vietnam to have the distinction of having the fastest-growing carbon dioxide emissions over that same time period. Emissions increased by 137 percent, exceeding the carbon-dioxide emissions growth in China, where emissions increased by 123 percent.[26]

Number of Days Needed to Consume 100 Kilowatt-hours

	Days to use 100 kWh
United States	3
Euro zone	5
South Africa	8
Tunisia	28
Ghana	106
Kenya	235
Nigeria	245
Tanzania	397
Ethiopia	702

In 2013, Todd Moss of the Center for Global Development came up with a clever way to illustrate the energy poverty that prevails in the developing world by calculating how long it takes residents of different countries to consume 100 kilowatt-hours of electricity. His findings: an average American uses that amount of power in about **66 hours**. By contrast, it takes the average Tanzanian more than a year to use that quantity of electricity. *Source*: Center for Global Development.[27]

So what can the United States and other wealthy countries say to a country like Vietnam when it comes to carbon dioxide emissions? The answer: very little. After all, Vietnam has about ninety million residents who live on an average annual income of about $1,400.

Or consider China. In 2011 alone, that country added 55,000 megawatts of new coal-fired capacity.[28] That capacity was equal to about three-quarters of all of the nuclear reactors under construction around the world in 2013. (That amount totaled 72,000 megawatts.)[29] Between 2012 and 2014, China was expected to add another 70,000 megawatts of new coal-fired electric-generation capacity, and another 280,000 megawatts of coal plants are being planned.[30] For comparison, in 2011, the United States had about 300,000 megawatts of coal-fired electric capacity.[31]

Proponents of renewable energy are quick to claim that countries like China and India should be moving more quickly toward renewables in general and wind energy in particular. But in his 2012 book, *The Carbon Crunch: How We're Getting Climate Change Wrong—and How to Fix It,* Dieter Helm, a professor of energy policy at Oxford University, points out that *just to match the growth* in coal-fired electricity in China and India, let alone replace any current power plants, those two countries would have to deploy about 1,000 megawatts of new wind turbines *every week* or about 52,000 per year. "Whether these numbers are out by a few thousand a year either way . . . does not much matter; it is the sheer scale of what is going on that matters."[32]

But we needn't look only at developing countries to see the essentiality of coal. Germany, which has the biggest economy in Europe, is rushing to close its nuclear reactors in the wake of the accident at the Fukushima facility in Japan. The result: German utilities are placing their bets on coal. By 2013, Germany was building 11,000 megawatts of new coal-fired power plants.[33]

The essentiality of electricity cannot be disputed. And that essentiality explains coal's centuries-long persistence in our fuel mix. We want electricity. We want it at home. We want it in our tents when we go camping. Electricity is so flexible and useful we want to be able to use it always, everywhere.

In short, we want batteries.

GOOG < Coal

The headline in the *New York Times* nearly gushed over the prospect: "A Subsidiary Charts Google's Next Frontier: Renewable Energy." The story, published November 28, 2007, explained that Google, a company with "a seemingly limitless source of revenue, plans to get into the business of finding limitless sources of energy."[34] Google, it said, was planning to create "renewable energy technologies that are cheaper than coal-generated power."

Google, which at the time was only nine years old and had almost no experience in the energy sector, did nothing to tamp down the media's expectations.[35] The Internet search and advertising company declared that it was planning "to spend tens of millions on research and development and related investments in renewable energy. As part of its capital planning process, the company also anticipates investing hundreds of millions of dollars in breakthrough renewable-energy projects." Larry Page, Google's cofounder, predicted that the company would be able to produce 1,000 megawatts of renewable energy capacity that is Cheaper than coal, and he added, "We are optimistic this can be done in years, not decades."[36] The company even created an equation that touted their goal: RE < C, or renewable energy Cheaper than coal.

That equation—and the optimism—didn't last very long. On November 22, 2011, less than four years after Google announced the RE < C effort, one of the company's bloggers quietly announced that the Cheaper-than-coal effort was being abandoned and that "other institutions are better positioned than Google to take this research to the next level."[37] When I asked a former Google employee who had been involved in the project why the company had abandoned RE < C, he admitted that the company didn't appreciate how long it takes to commercialize energy technologies. A new software program could be put together in a few months and launched almost immediately, he told me. Getting new energy systems into place takes years, and yes, even decades. Once the honchos at Google realized that, the company bailed out on RE < C.

AQUION ENERGY

Official Name Aquion Energy

Web site http://www.aquionenergy.com

Ownership Privately held

Headquarters Pittsburgh, PA

Finances By early 2013, it had raised more than $110 million in private equity and debt

2011 Revenue N/A

To make batteries dirt cheap, Jay Whitacre realized he was going to have to use materials that were almost as cheap as dirt.

To be sure, it doesn't take a rocket scientist to understand that if you want to mass-produce anything at low cost, you must use cheap materials. But Whitacre is, in fact, a rocket scientist. He used to work at the Jet Propulsion Laboratory. When he embarked on his quest to make batteries that were cheap, stable, durable, and environmentally friendly, he had to forget traditional battery chemistries.[38] Out went the usual suspects like lead, nickel, lithium, and cadmium. When he began testing various substances in his laboratory at Carnegie Mellon University in Pittsburgh, Whitacre set an almost absurdly low price threshold: none of his battery's ingredients could cost more than $2 per kilogram.[39]

Those limits left Whitacre with just a handful of options. But by trading relatively expensive elements for dirt-cheap substances like water, cotton, salt, charcoal, and manganese—and by surrendering some ground on energy density—Whitacre has

September 18, 2012: Battery designer Jay Whitacre at Aquion Energy's Pittsburgh facility. The gray chest he's standing behind will hold about 500 or 600 watt-hours of electricity and is designed for use in remote areas to power devices like vaccine refrigerators. *Source:* Photo by author.

come up with what may be the most important development in batteries since Edison introduced the nickel–iron–alkaline battery in 1909.[40]

Today, global spending on batteries totals about $56 billion per year.[41] While the battery market is huge, and growing, Whitacre's design, which he and his colleagues at Aquion Energy call an aqueous hybrid–ion cell, won't ever be used to power your phone, your car, or your electric drill. It's too heavy and bulky for those applications. Nor will it be deployed in a significant way in the next few years in Europe or the United States, where electricity is cheap and abundant. Nevertheless, Whitacre's new battery will find plenty of customers.

Consider the history and chemistry of batteries, which have long been the energy sector's killer app. Ever since the days of Alessandro Volta, who invented the battery in 1800, humans have been trying to find the best way to, in effect, put lightning in a bottle.[42] Coal can be stockpiled nearly anywhere. Oil can be poured into low–cost tanks. Natural gas can be compressed or liquefied for storage. But storing electrons has always been the prize. No other form of energy is as valuable or flexible as electricity. But batteries stink. They always have. They are uniformly too heavy, too bulky, and most important, they're too finicky. Just ask Boeing or Sony.

Aviation giant Boeing decided to use lithium–ion batteries in its 787 Dreamliner because of their high energy density. A lithium–ion battery can store about 150 watt–hours per kilogram, which is about two times as much as a typical nickel–cadmium and three times as much as a lead–acid battery.[43] But Boeing's decision to use lithium–ion batteries came with a heavy price. In January 2013, after a pair of 787s had onboard battery fires, the new jetliner was grounded for about four months.[44] The resulting retrofits to the 787, as well as the reputational loss, cost Boeing tens of millions of dollars. One Boeing customer sought some $37 million in compensation from the company.[45] In the wake of Boeing's problems, other airplane makers, including Airbus, Bombardier, and Mitsubishi, opted to use nickel-cadmium batteries in their new airplanes rather than risk using lithium-ion.[46]

Lithium-ion batteries dominate the world of consumer electronics. But few consumers enjoy seeing their laptop computer go up in flames. Unfortunately, that's exactly what happened a few years ago when a rash of laptop-battery fires forced Sony, the huge Japanese electronics firm, to recall about seven million lithium-ion batteries it had manufactured.[47]

Lithium-ion batteries have relatively high energy density, but they also degrade quickly in high temperatures. If they get fully discharged, they are ruined. And as Boeing and Sony have painfully learned, if the batteries are overcharged, or discharged too rapidly, they can catch fire.

Other battery chemistries can also present difficulties. Lead-acid batteries are relatively cheap and have proven their utility over the last 150 years. (The design was invented in 1859 by the French physician Gaston Planté.)[48] But lead has some serious downsides; it's a potent neurotoxin and it's heavy. Add in some sulfuric acid—used in the battery's electrolyte—and the design's environmental problems become apparent. Plus, lead-acid batteries can explode or catch fire. That's what happened to a large lead-acid storage system in Hawaii. In August 2012, a bank of batteries located near a 15-megawatt wind project in Hawaii went up in flames. Although firefighters used about 1,000 pounds of chemicals to try to douse the fire, they were eventually forced to just let the batteries burn.[49]

Sodium-sulfur batteries, which are used to store electricity for use on the electric grid, are also finicky. Although sodium-sulfur systems have been installed in more than 170 locations in six countries, the batteries operate at high temperature, which, by itself, makes them risky. In September 2011, a bank of sodium-sulfur batteries caught fire at a factory in Japan's Ibaraki prefecture owned by Mitsubishi Materials Corp. The blaze burned for more than two weeks before it was finally extinguished.[50]

As any chemist will tell you, the higher a substance's energy density, the more reactive it is.[51] A bucket filled with gasoline has very high energy density and can thus be made to ignite, or explode, with a small spark. A similar bucket filled with leaves or sawdust can also be set aflame, but the risk of explosion is almost nil, and any fire that starts will be far less dramatic than the one fed by a similar volume of gasoline.

When he began working on a new battery design, Whitacre, who has a PhD in materials science and engineering, understood the tradeoffs. If he pursued a battery with high energy density, it would tend to be unstable. Therein lies the genius of Whitacre's design: his battery is made largely of salt water, which is stable, heavy, and thermally resistant. It's hard to make water burn. The other feature (which some might call a flaw) of his design: it's big and heavy. For decades, battery designers have worked to make batteries Smaller and Smaller, and in doing so, increase their energy density. That's fine. But batteries work by getting ions to move from the anode to the cathode and back again. When you move lots of ions from one pole to the other, it can cause expansion and contraction as they enter the battery's electrodes. The more ions that move from anode to cathode and back, the greater the energy density of the battery. As that energy density increases, the more problematic the swelling and shrinking becomes.

Rather than see weight and low energy density as a hindrance, Whitacre used them to his advantage. In September 2012, when I visited Aquion's research and product-development facility near downtown Pittsburgh, Whitacre explained that by "having more mass, we are protecting ourselves from thermal swings." He went on, explaining that his design "traded high energy density for durability and low cost." The result: Whitacre's battery is a relative giant when compared to those used in hybrid vehicles or all-electric cars.

Aquion Energy's basic cell is called the AE12. A bank of eighty-four of Aquion's cells fits on a pallet weighing 2,750 pounds (1,250 kilograms). That pallet-load of batteries holds about 12 kilowatt-hours of energy.[52] By comparison, the lithium-ion battery pack in the all-electric Nissan Leaf automobile stores twice as much energy (24 kilowatt-hours) in a package that weighs about a quarter as much, 660 pounds (300 kilograms).[53] Put another way, the energy density of the Aquion battery is less than 10 watt-hours per kilogram. The Leaf's battery pack has eight times more energy density, 80 watt-hours per kilogram.[54]

But Whitacre doesn't care about the automobile market. His goal is to make big batteries that are Cheaper and more durable for the stationary market. In one of the test labs, Whitacre showed me a gray plastic chest, which was about the size of a large kitchen trash can. The chest was stuffed with cells, a neatly organized set of wires, and a power management system. "We can attach this to a small photovoltaic system, and in a day or so we can hook it up to a vaccine refrigerator. We can store enough electricity in this chest, maybe 500 or 600 watt-hours, to keep that vaccine refrigerator running for days. Those things are very efficient," Whitacre told me. "I was just explaining this to Bill Gates the other day."

Wait, what? Yes, Whitacre had a meeting with the Microsoft billionaire and philanthropist to talk about Aquion's progress. Such is the life of Whitacre, who might be the hottest property in the battery world. He's not a rock star or a billionaire. Not yet. But by early 2013, Aquion Energy had raised more than $110 million, including a significant chunk of cash from the venture capital firm Kleiner Perkins Caufield Byers, where Al Gore is a partner.[55] Whitacre's chat with Gates must have been convincing. In April 2013, Gates was among a group of investors who put $35 million into Aquion. (Gates has also invested in other energy storage companies, including LightSail Energy and Ambri.)[56]

Although Whitacre and Aquion have gained plenty of exposure and raised a large amount of money, the company isn't home free, not by a long shot. Moving from the laboratory to mass production is always fraught with challenges. In late 2013, Aquion began large-scale manufacturing of its batteries at a factory outside of Pittsburgh that was formerly used to build televisions. The company's goal is to produce batteries with a total capacity of 500 megawatt-hours per year.[57] That sounds like a lot. But keep in mind the enormous scale of global electricity demand and the minuscule amount of electricity that we are currently able to store.

If you were somehow able to collect all of the world's car batteries and string them together, you'd only have enough storage for about 1 terawatt-hour, or 1 trillion watt-hours.[58] In 2012, global electricity generation was about 22,500 terawatt-hours.[59] In other words, all of the world's automotive batteries combined can store only enough electricity to power the globe for about thirty minutes.[60]

The potential rewards for a company that can build a Cheaper battery are enormous. In 2012 alone, the value of global electricity sold was approximately $2.2 trillion.[61] A Cheaper, more durable electricity storage system would help make that trillion-dollar system work better. A Cheaper battery would help turn the intermittent energy that is produced by wind turbines and solar cells into more reliable power. A Cheaper battery would allow conventional electricity producers to store some of the energy they produce at night, when demand is low, and sell it during the day when demand is high. Such a system would reduce fuel costs and wear and tear on generation facilities and would be worth tens of billions of dollars per year to electricity providers all over the globe. It would also be attractive to precision manufacturers and other operations that demand highly reliable supplies of electricity.

Whitacre and his colleagues at Aquion believe their biggest near-term opportunity is in remote and island economies where electricity costs are high because they have to rely on diesel-fired generators for electricity. By combining battery storage with the diesel units and solar-photovoltaic systems, Aquion could dramatically reduce the cost of electricity in those regions. Another possible application: cell-phone towers, which need to continue operating even when the electric grid falters.

By marketing a battery that is safe, durable, and contains less toxic materials, Aquion will likely have an advantage over competitors. But the key advantage, of course, is that their battery should be Cheaper. And Cheaper always counts.

Ending this section of the book with a profile of Aquion and Jay Whitacre is appropriate because the company and the scientist are fine examples of American innovation and entrepreneurship in the energy sector. Energy policy and innovation are the primary themes of the next section, which also explains why the United States is particularly well positioned to dominate the Smaller Faster future.

PART IV:

Embracing Our
Smaller Faster Future

20

GETTING ENERGY POLICY RIGHT

Everything in our society—in fact, everything that happens inside of us—begins with the transformation of energy. No symphonies can be imagined or played, no planes can take off, no crops can be harvested without some form of energy being transformed into another.

We turn the food energy in tortillas into the sugars our muscles need to play a Woody Guthrie tune on the guitar. Internal-combustion engines convert the chemical energy in diesel fuel into the motive power that delivers more tortillas to the supermarket. We convert the photonic energy that hits our solar panels into electricity that feeds the lamps that illuminate our street signs.

Energy is the master resource. Therefore, we must make certain that our energy policies are in line with the trend toward Smaller Faster Lighter Denser Cheaper. Policies that promote low-density, expensive energy are destined to fail because they ignore both physics and economics. For decades, the catastrophists have been claiming that our future lies with renewable energy. In doing so, they have been supporting the increased use of sources that have fatally low power density: wind and biofuels.

REJECT WIND AND BIOFUELS

In promoting these subsidy-dependent sources, degrowthers have given momentum to landscape-destroying energy projects that can supply only a tiny fraction of the world's energy needs while doing next to nothing to reduce carbon dioxide emissions.

Over the past decade or so, I've written extensively about the problems with wind and biofuels. But since 2010, when I published *Power Hungry* (which cast a sharply critical eye on the foolishness of wind energy), and since 2008, when I published *Gusher of Lies* (which exposed the absurdities of the corn ethanol scam), those two forms of energy have continued to get both mandates and subsidies. The fundamental problem with both wind and biofuels is that they are not dense. Producing significant quantities of energy from either wind or biomass simply requires too much land. The problem is not one of religious belief, it's simple math and basic physics.

Energy is the pillar upon which economic growth is built. We need to pull the blinkers from our eyes when it comes to energy and recognize that cheap, abundant, reliable energy creates wealth. If we want to grow our economy, we cannot rely on the mirage of wind and biofuels.

WIND ENERGY'S INCURABLE DENSITY PROBLEM

The next time you read an article stating, or hear a pundit claim, that we can run the world using wind energy, remember this figure: 1 watt per square meter. That's the power density of wind energy.

The punch line is this: even if we ignore wind energy's incurable intermittency, its deleterious impact on wildlife, and how 500-foot-high wind turbines blight the landscape and harm the landowners who live next to them, its paltry power density simply makes it unworkable. Wind-energy projects require too much land and too much airspace. In the effort to turn the low power density of the wind into electricity, wind turbines standing about 150 meters high must sweep huge expanses of air.[1] (A 6-megawatt offshore turbine built by Siemens sports turbine blades with a total diameter of 154 meters that sweep an area of 18,600 square meters.[2] That sweep area is nearly three times the area of a regulation soccer pitch.[3])

By sweeping those enormous expanses of air, wind turbines are killing large numbers of bats and birds. A 2013 peer-reviewed study estimated that wind turbines in the United States are killing nearly

900,000 bats and 573,000 birds per year, including some 83,000 birds of prey.[4] Another 2013 study, done by some of the US Fish and Wildlife Service's top raptor biologists, found that the number of eagles killed by wind turbines has skyrocketed, and that increase has occurred alongside the rapid increase in wind-energy capacity. In 2007, the United States had 17,000 megawatts of installed wind capacity.[5] That year, the biologists were able to verify two eagle kills by turbines. By 2011, installed wind capacity had nearly tripled to 47,000 megawatts, and the number of verified eagle kills by wind turbines had increased to 24.[6] Thus, over a time period when wind capacity tripled, the number of eagle kills increased twelvefold. Between 1997 and mid-2012, at least eighty-five eagles, including six bald eagles and seventy-nine golden eagles, were killed by wind turbines. That tally did not include the ongoing eagle slaughter at California's Altamont Pass, where about one hundred golden eagles are killed by turbines every year.[7]

Every one of those kills was a violation of the Bald and Golden Eagle Protection Act. But it wasn't until late 2013 that the wind industry was finally brought to justice. On November 22, 2013, the Justice Department announced that it had reached a $1 million settlement with Duke Energy. Duke pled guilty to criminal violations of the Migratory Bird Treaty Act for killing 14 golden eagles and 149 other protected birds at two company-owned wind projects in Wyoming.[8] In addition to the wildlife toll, wind turbines create audible noise as well as low-frequency noise and infrasound that is injurious to human health. Although the wind industry continues to deny the existence of a problem, numerous studies, as well as a wealth of news clippings from around the world, show that the noise problem with wind turbines is real and widespread. (See Appendix E.)

There are hundreds of examples of the growing global backlash against Big Wind. To cite just one: in July 2013, more than 2,000 protesters marched in Ireland to oppose a wind-energy project that could result in the installation of more than a thousand wind turbines in that country's midlands region.[9] From Ireland to New Zealand and Massachusetts to Wisconsin, there is growing outrage among rural and semi-rural homeowners about the encroachment of massive wind projects. The European Platform Against Windfarms now lists some six hundred signatory organizations from twenty-four countries.[10] In the UK—where fights are raging against industrial wind projects in Wales, Scotland, and elsewhere—some three hundred anti-wind groups have been formed.[11] Meanwhile, here in the United States, about 150 anti-wind groups are active.[12] In Ontario, Canada, the epicenter of the backlash against Big Wind, there are fifty-five anti-wind groups.[13]

Big Wind has become so concerned about the backlash that it has taken to suing opponents. In Ontario, NextEra Energy, the Florida-based electricity provider, filed a SLAPP (strategic lawsuit against public participation) suit against Esther Wrightman, a resident of tiny Kerwood, Ontario. The suit, filed on May 1, 2013, is a blatant effort to shut Wrightman up. NextEra, which has a market capitalization of $35 billion, claims that Wrightman, a thirty-two-year-old married mother of two young children, who has effectively no assets, misused its logo and libeled it by calling the company "NexTerror." Wrightman is now having to defend herself in court (she can't afford to hire a lawyer of her own) against some of Toronto's biggest law firms.[14] Of course, you won't read about the backlash against Big Wind in the *New York Times* or leftist publications like *Mother Jones*. But the backlash is strong and it's growing.

The bird-and-bat kills, noise problems, and growing backlash are all products of wind energy's low power density. Over the past seven years or so, I've been collecting power-density estimates from a variety of energy analysts, including Jesse Ausubel, David J. C. MacKay, Vaclav Smil, Todd A. Kiefer, and David Keith. In addition, I have collected my own data on wind projects around the world. The result: the power density of wind energy is 1 watt per square meter. Period. End of story. Elvis has left the building.

Let's start by looking at Ausubel, the director of the Program for the Human Environment at Rockefeller University. Ausubel has been writing about energy for decades. In 2007, he published "Renewable and Nuclear Heresies," a paper in which he put wind energy's power density at 1.2 watts per square meter.[15]

Wind energy was considered modern back in 1872, in East Hampton, Suffolk County, New York. *Source*: Library of Congress, HAER NY, 52-HAMTE, 2—34.

In 2009, MacKay, an engineering professor at the University of Cambridge, published *Sustainable Energy—Without the Hot Air,* in which he debunked many of the myths about wind energy and other renewable sources. MacKay put the areal power density of onshore wind at 2 watts per square meter and offshore wind at about 3 watts per square meter.[16] But that latter figure is of little importance, as very little offshore wind capacity has been (or will be) built. The reason: offshore wind-energy projects cost three to four times as much as onshore ones.[17]

Now consider Smil, a geographer at the University of Manitoba and author of more than three dozen books. In 2010, he wrote what he calls a "power density primer" in which he laid out the energy flows that could be harnessed from various sources. His finding: wind energy's power density ranges 0.5 to 1.5 watts per square meter.[18]

T. A. "Ike" Kiefer, a recently retired Navy captain, aviator, and lecturer at the US Air Force Air War College, has done an exhaustive analysis of various renewable-energy sources with a special emphasis on biofuels. Kiefer studied wind energy data collected by the National Renewable Energy Laboratory on projects located across the United States from 2000 to 2009. In January 2013, Kiefer published "Twenty-First Century Snake Oil: Why the United States Should Reject Bio-fuels as Part of a Rational National Security Energy Strategy." Kiefer's finding: US wind energy has an areal power density of 1.13 watts per square meter.[19]

In February 2013, Amanda S. Adams, a geoscientist who studies mesoscale atmospheric modeling at the University of North Carolina at Charlotte, and David W. Keith, an applied physicist at Harvard, published a paper in *Environmental Research Letters* titled "Are Global Wind Power Resource Estimates Overstated?" Adams and Keith wrote that several estimates "have assumed that wind power production of 2 to 4 watts per square meter can be sustained over large areas." But in their analysis, Adams and Keith found that "wind shadow" occurs when wind turbines are put in large arrays.[20] The shadowing effect reduces the output of the turbines and, in doing so, requires them to be spread farther apart. The conclusion of their study: "It will be difficult to attain large-scale wind power production with a power density of much greater than 1.2 watts per square meter contradicting the assumptions in common estimates of global wind power capacity." And for very large wind projects, those larger than about 100 square kilometers, the power density of wind energy "is limited to about 1 watt per square meter."[21]

Finally, in my own research, I collected data on sixteen different projects that ranged in size from 40 megawatts to more than 2,000 megawatts. The projects were geographically diverse—Texas, Pennsylvania, Wyoming, Kansas, Ontario, and Australia—and totaled more than 5,000 megawatts of capacity. (See Appendix F.) My finding: the average power density was 2.29 watts per square meter.[22] However, that number must be adjusted because the wind doesn't blow all the time. Wind projects generally have capacity factors of anywhere from 20 to 40 percent.[23] "Capacity factor" is the term used for the amount of time that an electric generator produces at full power. Thus, if we assume a (generous) 39 percent capacity factor for wind projects and multiply that by my sixteen-project average of 2.3 watts per square meter (0.39 x 2.3), then the areal power density of wind energy is 0.9 watts per square meter.

In summary, the power-density calculations of wind energy are as follows:

> Jesse Ausubel: 1.2
> David J. C. MacKay: 2
> Vaclav Smil: 1
> Todd Kiefer: 1.13
> Adams/Keith: 1
> Robert Bryce: 0.9

Add those figures together and divide by six, and you get an average power density for wind energy of 1.2 watts per square meter—exactly what I reported in *Power Hungry*. If we toss out the high and low estimates (MacKay's 2 watts per square meter, and my 0.9 watts per square meter), then the average areal power density is 1.05 watts per square meter.

The simple, undeniable truth is that wind energy cannot, *will not*, become a major supplier of energy in the United States or other countries because it requires too much land. Proving that point requires only that we use our 1 watt per square meter metric and apply it to the domestic coal industry. In 2011, the United States had about 300 gigawatts (300 billion watts) of coal-fired generation capacity.[24] If policy makers wanted to replace all that coal-fired capacity with wind turbines—at

If You Want to Replace US Coal-fired Capacity with Wind, Then Find a Land Area the Size of Italy

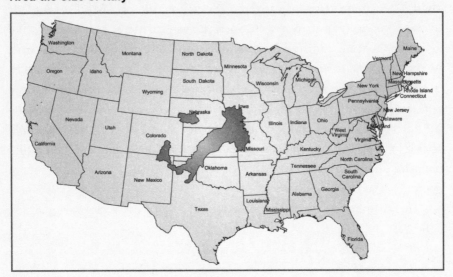

With an areal power density of 1 watt per square meter, wind energy requires vast tracts of land. Replacing America's coal-fired electricity-generation capacity, which is roughly 300 billion watts, would require some 300 billion square meters, a land area the size of Italy. *Source*: Author calculations, based on power density calculations for wind published by Jesse Ausubel, Vaclav Smil, David J. C. MacKay, Todd Kiefer, Amanda S. Adams, and David W. Keith.

1 watt per square meter—they would need to set aside a land area of 300 billion square meters, or 300,000 square kilometers, or roughly 116,000 square miles. That's a land area the size of Italy.

If that concept isn't silly enough, here's the kicker: because of the noise produced by the turbines, no one could live on that land. You are unlikely to hear about the problem with wind-turbine noise in the mainstream media. The wind industry's many apologists are eager to dismiss the noise complaints. But the problem is real. Residents who've been forced to flee from their homes after wind turbines were built near them—and I've personally interviewed many of these people—are consistent in their complaints of headaches, nausea, sleeplessness, and other symptoms caused by the turbines.

Regardless of whether the issue is human impacts or wildlife impacts, *the energy sprawl that comes with wind-energy projects makes it unsustainable.* Dieter Helm neatly summed up that point in his book *The Carbon Crunch.* "Even if we devoted all our resources to current wind and solar technologies, they would not be anything like enough to solve the problem of climate change. There simply is not enough land."[25] Exactly right. There simply is not enough land to make wind energy viable on a global basis, or in many cases, even on a regional basis.

In 2011, James Hansen, one of the world's most famous climate scientists, delivered a similar rebuke to the environmentalists who've been touting wind and solar as the solutions to our energy needs. In a posting on his blog, Hansen wrote that "suggesting that renewables will let us phase rapidly off fossil fuels in the United States, China, India, or the world as a whole is almost the equivalent of believing in the Easter Bunny and Tooth Fairy." He went on to say that politicians and environmental groups "pay homage to the Easter Bunny fantasy, because it is the easy thing to do . . . They are reluctant to explain what is actually needed to phase out our need for fossil fuels."[26]

In late 2013, Hansen and three other climate scientists wrote an open letter to environmentalists encouraging them to support nuclear. They stated: "Continued opposition to nuclear power threatens humanity's ability to avoid dangerous climate change." They further stated, "Renewables like wind and solar and biomass will certainly play roles in a future energy economy, but those energy sources cannot scale up fast enough to deliver cheap and reliable power at the scale the global economy requires."[27]

Even if renewable sources like wind and solar were subsidized to the point where they were providing significant amounts of our energy needs, we would still need to back up those sources with conventional generation units fired by coal or natural gas.

Like them or not, hydrocarbons and nuclear are Denser sources than renewables. They allow us to produce large amounts of energy from relatively small pieces of land. Density is green. We humans should be striving for density in nearly everything we do—from energy production and food production to computers and buildings. Wind energy is not dense. Therefore it is not green.

Debunking the Big Fibs About Wind and Solar

For decades, major environmental groups have been making many false claims that renewable energy can meet all of humankind's needs and provide big cuts in global carbon dioxide emissions. For example, Greenpeace (which has an annual operating budget of about $300 million) claims that renewable energy, "smartly used, can and will meet our demands. No oil spills, no climate change, no radiation danger, no nuclear waste."[28]

The Sierra Club (2012 budget: $100 million), shares Greenpeace's disgust with hydrocarbons and nuclear. The club has a "beyond oil" campaign, a "beyond coal" campaign, and a "beyond natural gas" campaign. The group says, "we have the means to reverse global warming and create a clean, renewable energy future."[29]

In June 2012, the *New York Times* ran an op-ed written by a Swedish academic, Christian Azar, and two economists from the Environmental Defense Fund (2012 revenue: $116.5 million), Thomas Sterner and Gernot Wagner, which said the world needs to "kick its addiction to fossil fuels" and that "the solar and wind revolution is just beginning."[30]

In 2013, Bob Deans, the associate director of communications for the Natural Resources Defense Council (2012 revenue: $103 million), declared that "we need to turn away from the fossil fuels of the past, invest in efficiency and renewables and build a twenty-first century economy on new fuels . . ."[31]

In 2013, Josh Fox, the activist and director of the anti–natural gas film *GasLand,* said, "renewable energy solutions exist, right now. We are facing a real choice here and we must move towards non-fossil-fuel sources . . . the future is renewable energy." Fox went on, "Renewables can do the job, we don't need nuclear to solve the problem."[32]

Those claims are easy to make. But they are even easier to debunk. We can do so by ignoring the need for oil (which is used primarily for transportation) and focusing solely on global electricity demand.

Between 1985 and 2012, global electricity production increased by about 450 terawatt-hours per year.[33] That's the equivalent of adding about one Brazil (which used 554 terawatt-hours of electricity in 2012) to the global electricity sector every year.[34] And the International Energy Agency expects global electricity use to continue growing by about one Brazil per year through 2035.[35]

What would it take to just keep up with the *growth* in global electricity demand—450 terawatt-hours per year—by using solar? We can answer that question by looking at Germany, which has more installed solar-energy capacity than any other country, about 33,000 megawatts.[36] In 2012, Germany's solar facilities produced 28 terawatt-hours of electricity.[37] Thus, just to keep pace with the growth in global electricity demand, the world would have to install 16 times as much photovoltaic capacity as Germany's entire installed base, *and it would have to do so every year.*

Prefer to use wind? Fine. In 2012, the world's wind turbines—some 284,000 megawatts of capacity—produced 521 terawatt hours of electricity. The United States has more wind capacity than any other country, about 60,000 megawatts at the end of 2012.[38] Thus, just to keep pace with electricity demand growth, the world would have to install about four times as much wind-energy capacity as the United States has right now, *and it would have to do so annually.* That is simply not going to happen. Gritty activists like Esther Wrightman and other rural residents who value their homes and communities won't let it happen.

The point here is so obvious that even Ray Charles could have seen it: solar energy and wind energy cannot even keep pace with the growth in global electricity demand, much less displace significant quantities of hydrocarbons.

Now let's look at carbon dioxide emissions. The American Wind Energy Association claims that wind energy reduced US carbon dioxide emissions by 80 million tons in 2012.[39] That sounds significant. But consider this: global emissions of that gas totaled 34.5 billion tons in 2012. Thus, the 60,000 megawatts of installed wind-generation capacity in the United States reduced global carbon dioxide emissions by about two-tenths of 1 percent. That's a fart in a hurricane.

To make the point even clearer, let's look at the history of global carbon dioxide emissions. Since 1982, carbon dioxide emissions have been increasing by an average of about 500 million tons per year.[40] If we take the American Wind Energy Association's claim that 60,000 megawatts of wind-energy capacity can reduce carbon dioxide emissions by 80 million tons per year, then simple math shows that if we wanted to stop the growth in global carbon dioxide emissions by using wind energy alone—and remember doing so won't reduce any of the existing demand for coal, oil, and natural gas—we would have to install about 375,000 megawatts of new wind-energy capacity every year.

How much land would all those wind turbines require? Recall that the power density of wind energy is 1 watt per square meter. Therefore, merely halting the rate of growth in carbon dioxide emissions with wind energy—and remember this would not displace any of our existing need for coal, oil, and natural gas—would require covering a land area of about 375 billion square meters or 375,000 square kilometers. That's an area the size of Germany. *And we would have to keep covering that Germany-sized piece of territory with wind turbines every year.*

What would that mean on a daily basis? Using wind to stop the growth in carbon dioxide emissions would require us to cover about 1,000 square kilometers with wind turbines—a land area about 17 times the size of Manhattan Island—and *we would have to do so every day.*[41] Given the ongoing backlash against the wind, the silliness of such a proposal is obvious.

The hard reality is that wind turbines are nothing more than climate-change scarecrows.

Over the past few years, the United States and other countries have been subsidizing the paving of vast areas of the countryside with 500-foot-high bird-and-bat-killing whirligigs that are nothing more than climate talismans.[42] Wind turbines are not going to stop changes in the Earth's climate. Instead, they are token gestures—giant steel scarecrows—that are deceiving the public into thinking that we as a society are doing something to avert the possibility of catastrophic climate change.

BIOFUELS ARE "A CRIME AGAINST HUMANITY"

Many critiques have been written about the foolishness of America's mandates and subsidies for biofuels. But the most savage was almost certainly published in March 2013 by Ike Kiefer, who launched this barrage:

Imagine if the US military developed a weapon that could threaten millions around the world with hunger, accelerate global warming, incite widespread instability and revolution, provide our competitors and enemies with cheaper energy, and reduce America's economy to a permanent state of recession. What would be the sense and the morality of employing such a weapon? We are already building that weapon—it is our biofuels program.[43]

Ouch.

Remember, this guy's on our side. Kiefer represents the finest intellectual tradition in the US military. He has a bachelor's degree in physics from the US Naval Academy and a master's in strategy from the US Army Command and General Staff College. He's also a warrior. Kiefer was deployed seven times and spent twenty-one months in Iraq. In his scathing indictment of the biofuels sector, which was published in *Strategic Studies Quarterly*, the US Air Force's most prestigious journal, Kiefer declares, "For the sake of our national energy strategy and global security, we must face the sober facts and reject biofuels while advocating an overall national energy strategy compatible with the laws of chemistry, physics, biology, and economics."[44]

To all of that, I say Amen, Hallelujah, and Pass the Biscuits. What makes Kiefer's massively footnoted takedown of biofuels so effective is that he doesn't frame his argument against biofuels with moral cries about higher food prices, even though that's one of his key points. Instead, he hammers the physics and math. And in particular, he focuses on density.

Kiefer (who retired from the Navy in mid-2013) explained that biofuels have "an anemic power density of only 0.3 watts per square meter." For comparison, modern solar photovoltaic panels are about 6 watts per square meter, or 20 times more; an average oil well producing 10 barrels per day is 27 watts per square meter, and an average nuclear plant is more than 50 watts per square meter.[45]

The low areal power density of biofuels cannot be overcome. That low power density is due to the limits of photosynthesis, and despite decades of trying, we haven't been able to improve on it. Chlorophyll remains the preeminent converter of sunlight into energy, but it does so at its own pace.

The low power density of biofuels means that vast amounts of land are needed to produce even small quantities of fuel. For example, if we wanted to replace all of the oil used for transportation in the United States with corn-based ethanol, writes Kiefer, it would require about 700 million acres of land to be planted in nothing but corn. That would be "37 percent of the total area of the continental United States, more than all 565 million acres of forest, and more than triple the current amount of annually harvested cropland." Do you like biodiesel better? Kiefer calculates that relying on soy biodiesel to replace domestic oil needs would "require 3.2 billion acres—one billion more than all US territory including Alaska."[46] Kiefer's thirty-eight-page paper includes more than one hundred footnotes and a half dozen charts or tables.

The Obama administration's response to Kiefer's assault consisted of two bland documents posted on *Strategic Studies Quarterly*'s Web site, neither of which addressed any of his arguments.[47] Rather than deal with critics like Kiefer, the Obama administration just keeps ladling out pork to the biofuel sector. Just two months after Kiefer's piece was published, the Obama administration announced—on a Friday afternoon, just before the 2013 Memorial Day weekend—that the Department of Defense was giving contracts worth a total of $16 million to three biofuel plants located in Illinois, Nebraska, and California.[48]

Kiefer's shootdown of biofuels was among a string of devastating investigations into the biofuels business published in recent years. Before looking at those, let's recall some of the whoppers that we've been told over the last few years about biofuels.

The biggest of them all was likely uttered in 2011, when President Barack Obama declared: "We can break our dependence on oil with biofuels."[49] That statement is consistent with Obama's political career, which has been notable for his unstinting support of biofuels. In 2006, Senator Obama, along with four other farm-state senators, sent a letter to President Bush asking him to ignore calls to reduce tariffs on Brazilian sugarcane-based ethanol. Lowering the tariff, they said, would make the United States dependent on foreign ethanol. "Our focus must be on building energy security through domestically produced renewable fuels," they wrote.[50] A few months later, Obama and two other farm-state senators introduced a bill that would promote the use of ethanol, mandate the use of more biodiesel, and create tax credits for cellulosic ethanol production. They called their bill the "American Fuels Act of 2007."

In 2008, after Obama was elected president, one of his first moves was to appoint former Iowa governor and longtime biofuels booster Tom Vilsack as his secretary of agriculture. Vilsack, Obama said, would be part of the "team we need" to strengthen rural America, create "green jobs," and "to free our nation from its dependence on oil."

In 2006, an arm of the Center for American Progress, a leftist think tank, launched a campaign called "Kick the Oil Habit."[51] The group's lead spokesman was actor Robert Redford, who appeared on TV talk shows and wrote opinion pieces in which he said America should quit using oil altogether so that it can get away from "dictators and despots." The Sundance Kid's solutions? Ethanol, biofuels, and hybrid vehicles.[52] That year, during an appearance on CNN's *Larry King Live*, Redford said that he supported corn ethanol production because "it's cheaper. It's cleaner. It's renewable. And you know what? It's American because we grow it."[53]

In a 2006 speech at the University of California–Los Angeles, former president Bill Clinton provided the Butch Cassidy to Redford's Sundance by repeating the spurious claim that oil is directly linked to terrorism. Clinton asked the students at UCLA: "Aren't you tired of financing both ends of the war on terror?" And he added that students should be worried that the money sent to oil-producing nations "might be diverted to destructive purposes."

Or consider Silicon Valley multimillionaire and venture capitalist Vinod Khosla. In 2006, at the same time that he was investing in various biofuel companies, Khosla claimed that making ethanol from cellulosic material was "brain-dead simple to do" and that commercial production of cellulosic ethanol was "just around the corner."[54] A few months later, Khosla was again hyping cellulosic ethanol, saying that biofuels could completely replace oil for transportation and that cellulosic ethanol would be cost competitive with corn-ethanol production. That same year, Khosla and former Senate minority leader Tom Daschle wrote an op-ed for the *New York Times* touting Brazil's "energy independence miracle." They said that Brazil proves that "an aggressive strategy of investing in petroleum substitutes like ethanol can end dependence on imported oil."[55]

In late 2011, one of Khosla's ventures, Range Fuels, a Georgia company that claimed it could profitably turn wood chips into ethanol, defaulted on an $80 million loan that was backed by the federal government. A few months later, Range, which had also received a $76 million grant from the Department of Energy, declared bankruptcy.[56] Khosla refused to comment on Range's bankruptcy.

Al Gore was yet another biofuel booster. After his 2006 movie, *An Inconvenient Truth*, was released, the former vice president promised that cellulosic ethanol would be "a huge new source of energy, particularly for the transportation sector. You're going to see it all over the place."[57]

Among the biggest—and most hawkish—promoters of biofuels has been James Woolsey, a former director of the Central Intelligence Agency. In 2007, during a speech to the Virginia Soybean, Corn and Grain Association, Woolsey said, "We must move from a hydrocarbon-based society to a carbohydrate-based society." Woolsey was among several defense hawks who frequently conflated oil and terrorism. During that same speech, Woolsey said, "The next time you pull into a gas station to fill your car with gas, bend down a little and take a glance in the side-door mirror. What you will see is a contributor to terrorism against the United States."[58]

Woolsey was part of a group called Set America Free, comprising a who's who of the military-agricultural-industrial complex. Created in 2004 and dominated by a group of neoconservatives who advocated for the Second Iraq War, the group frequently endorsed biofuels as a way toward realizing the delusional concept of "energy independence." In a 2008 editorial in the *Chicago Tribune,* one of the group's founders, Gal Luft, and his fellow traveler, Robert Zubrin, an author who advocates colonizing Mars, declared that "farm commodity prices have almost no effect on retail prices." The two concluded their May 6 screed by saying that the goal should be to "take down" the Organization of the Petroleum Exporting Countries and that "rather than shut down biofuel programs . . . we need to radically augment them."[59]

While Obama, other politicians, and actor/activists have been consistent—and consistently wrong—about the ability of biofuels to make a significant dent in our energy diet, the most persistent cheerleader for the biofuel insanity has been Amory Lovins, the cofounder of the Rocky Mountain Institute and a longtime ally of the bloggers at the Center for American Progress. In 1976, Lovins wrote an article for *Foreign Affairs* in which he claimed that "exciting developments in the conversion of agricultural, forestry and urban wastes to methanol and

other liquid and gaseous fuels now offer practical, economically interesting technologies sufficient to run an efficient US transport sector."
Given better efficiency in automobiles and a large enough installation
of cellulosic ethanol distilleries, he claimed that "the whole of the transport needs could be met by organic conversion."[60]

Ever since 1976, Lovins has been promoting biofuels. In 2011,
Lovins and a group of coauthors published *Reinventing Fire*. In an info-
graphic promoting the book, Lovins and his colleagues claim that by
2050, the United States will be getting 23 percent of its total energy
from "non-cropland biofuels."[61] This is ludicrous beyond language.

Proving that point requires only some elementary math. Let's assume that in 2050, the United States is still using about the same amount
of primary energy as it does today, which is about 45 million barrels
of oil equivalent per day.[62] Thus, Lovins claims that the United States
will be producing about 10.35 million barrels of oil equivalent per day—
which is 3.77 billion barrels per year—from biofuels by 2050.

Stick with me here as we walk through the numbers. Cellulosic
ethanol companies like Coskata and Syntec have claimed that they can
produce about 100 gallons of ethanol per ton of biomass. Recall that
100 gallons of ethanol is equal to about 66 gallons of gasoline, or about
1.57 barrels of oil. For simplicity, let's assume a ton of biomass can be
turned into 1.5 barrels of oil equivalent.

The next question is the productivity of the land. Oak Ridge National Laboratory says that an acre of switchgrass can produce 11.5 tons
of biomass per year.[63] Therefore, an acre of switchgrass can produce
17.25 barrels of oil equivalent per year. Remember that Lovins and his
coauthors claim we can produce 3.77 billion barrels of oil equivalent
from biofuels per year. Now we need to divide that quantity by the
17.25 barrels of oil equivalent per acre per year that we got from Oak
Ridge National Laboratory.

Simple division shows that the United States would need to set aside
about 219 million acres of land in order to grow enough biomass to
produce the amount of energy (10.35 million barrels of oil equivalent per
day) that Lovins and his pals are claiming in *Reinventing Fire*. That 219
million acres is roughly equal to 342,000 square miles or about 886,000

Amory Lovins's Vision for Biofuels: Producing 23 Percent of US Energy by 2050 from Plants Would Require Three Italys of Land

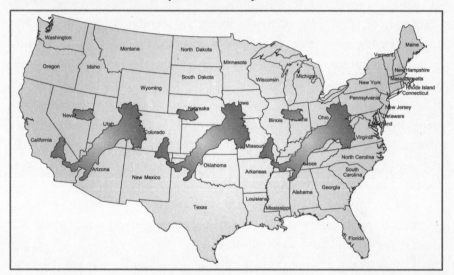

Source: Author calculations, based on land-productivity calculations for biofuels published by Oak Ridge National Laboratory.

square kilometers. That's a land area the size of Texas, New York, and Ohio, *combined*.[64] To produce enough biomass to meet Lovins's claim that biofuels will be supplying 23 percent of US energy by 2050 would require cultivating a land area nearly *three times the size of Italy*.

The absurdity of Lovins's claims are obvious. Yet, he's been lauded like few other Americans. In 2007, when I was working on a profile of Lovins, he volunteered his bio, which claimed that he has worked in more than fifty countries and that he has been awarded the "Blue Planet, Volvo, Onassis, Nissan, Shingo, and Mitchell Prizes; the Benjamin Franklin and Happold Medals; nine honorary doctorates, honorary membership in the American Institute of Architects, Life Fellowship of the Royal Society of Arts, and the Heinz, Lindbergh, *Time* Hero for the Planet, and World Technology Awards." He's also received a "genius" grant from the MacArthur Foundation. Despite those lofty credentials, it appears that Lovins has never bothered to use a calculator to see just how his biofuel plans pencil—or rather, how they don't.

Whether the salesman was Obama, Clinton, Gore, Woolsey, Khosla, or Lovins, one thing has become stunningly obvious: we've been had.

Biofuels, we were told, were the magic bullet, the energy-independence-punish-the-Arabs-anti-terror-better-than-standard-diesel-fuel miracle elixir. It wasn't true. It's never been true. Despite tens of billions in taxpayer money that have been thrown at corn ethanol, soy diesel, algae, and the rest, the US economy, and more particularly, the US military, has gained nothing from its biofuel dalliances. And it won't gain anything in the future because biofuels can't overcome basic math and physics. Nevertheless, the Obama administration continues to ladle subsidy gravy on nearly any hillbilly who can swear—cross-his-heart—that he has a recipe for spinning oil out of sawdust.

But enough about Lovins, Obama, and their ilk. Here are a few snippets from the more than two dozen studies done in recent years that have exposed the dark side of biofuels.

Let's start by looking at the US Environmental Protection Agency's own data, which shows that increased use of corn ethanol in gasoline means worse air quality. In 2007, the EPA determined that increased use of ethanol in gasoline would increase emissions of key air pollutants like volatile organic compounds and nitrogen oxide by as much as 7 percent. In a fact sheet regarding the Renewable Fuel Standard, the agency said, "Nationwide, EPA estimates an increase in total emissions of volatile organic compounds and nitrogen oxides (VOC + NOx) [of] between 41,000 and 83,000 tons." It went on, saying "areas that experience a substantial increase in ethanol may see an increase in VOC emissions between 4 and 5 percent and an increase in NOx emissions between 6 and 7 percent from gasoline powered vehicles and equipment."

The corn ethanol scam not only worsens air quality, it also makes food more expensive. In 2008, Mark W. Rosegrant of the International Food Policy Research Institute, a Washington, DC–based think tank whose vision is "a world free of hunger and malnutrition," testified before the US Senate on biofuels and grain prices.[65] Rosegrant said that the ethanol mandates caused the price of corn to increase by 39 percent, rice to increase by 21 percent, and wheat by 22 percent. He estimated that if the global biofuels mandates were eliminated altogether, corn prices would drop by 20 percent. Rosegrant added: "If the current biofuel expansion continues, calorie availability in developing countries is expected to grow more slowly; and the number of malnourished children is projected to increase."[66]

Also in 2008, an internal report by the World Bank found that grain prices increased by 140 percent between January 2002 and February 2008. "This increase was caused by a confluence of factors but the most important was the large increase in biofuels production in the US and EU. Without the increase in biofuels, global wheat and maize [corn] stocks would not have declined appreciably and price increases due to other factors would have been moderate."[67]

In 2011, Tim Searchinger, a research scholar at the Woodrow Wilson School at Princeton University, further clarified the link between biofuels and food prices.[68] In a June 16, 2011, article in *Scientific American*, Searchinger stated, "Since 2004[,] biofuels from crops have almost doubled the rate of growth in global demand for grain and sugar and pushed up the yearly growth in demand for vegetable oil by around 40 percent. Even cassava is edging out other crops in Thailand because China uses it to make ethanol."[69] The global push for biofuels, he wrote, requires us to consider the "moral weight" of what we are doing. "Our primary obligation is to feed the hungry. Biofuels are undermining our ability to do so. Governments can stop the recurring pattern of food crises by backing off their demands for ever more biofuels."[70]

In April 2013, the British think tank Chatham House released a blistering report on biofuels, which said the use of biofuels "increases the level and volatility of food prices, with detrimental impacts on the food security of low-income food-importing countries." It went on to say that due to land use changes, emissions from the production of biodiesel made from vegetable oils are "worse for the climate than fossil diesel." Finally, it says that the current 5 percent mandate for biofuel use in the UK will cost the country's motorists "in the region of $700 million" in 2013, and that the costs are likely to increase to $2 billion per year by 2020.[71]

At the outset of this chapter, I made it clear that I'm partial to Ike Kiefer's critique of biofuels. But when it comes to a guns-blazing, raging-with-fury analysis, few write-ups can match the one done by Jean Ziegler, a former member of the Swiss Parliament who served as the UN Special Rapporteur on the Right to Food from 2000 to 2008. In August 2013, Ziegler published *Betting on Famine: Why the World Still Goes Hungry*. (Disclosure: I read an advance copy of the book and provided a blurb for it.) Ziegler's book seethes with anger. He describes visits to biofuel plantations all over the world, and in each location he finds similar stories: exploited workers and expropriated land. Ziegler writes that the companies that produce biofuels have succeeded at convincing the public and politicians in Western countries that "energy from plant sources constitutes the miracle weapon against climate change. Yet their argument is a lie."[72]

Between 2006 and 2011, global biofuels production doubled. Ziegler reports that in 2011, global biofuels output totaled 600 million barrels, or about 1.64 million barrels per day. But ethanol contains only about two-thirds of the heat energy of oil. Therefore, the actual energy produced from biofuels in 2011 was closer to 1.2 million barrels of oil equivalent per day.[73] Ziegler reports that producing that volume of fuel required 100 million hectares of land.[74]

Let's put those numbers into perspective. Global energy use from all sources is currently about 250 million barrels of oil equivalent per day.[75] Therefore, biofuels are providing less than one-half of 1 percent of world energy needs. In doing so, they are requiring 100 million hectares (247 million acres).[76] That's a land area more than twice the size of California, or nearly twice the size of France.[77]

This is madness—the kind of madness that meets insanity coming the other direction. In order to produce about one-half of 1 percent of world energy needs, the biofuels sector is cultivating a land area nearly twice the size of France—or an area about half the size of all American cropland.[78] And yet, the president of the United States is claiming that biofuels can replace oil? It's difficult to imagine a bigger—let's call it what it is—lie.

Ziegler's blistering take on biofuels is equal parts travelogue and outraged screed. One of his stops is the sugarcane fields of Brazil, where he visits Father James Thorlby, better known as Father Tiago, a Scottish priest who has become an ardent defender of Brazil's exploited farm-workers. Ziegler doesn't mince words. After visiting the forlorn camps that house the workers who cut the sugarcane, which is used to produce ethanol for motor fuel, he writes that the workers have no choice but to accept "low wages, inhumane work hours, nearly nonexistent accommodations, and work conditions that approach slavery."[79] Ziegler visits other countries, including Colombia, Cameroon, and India, and in each location he finds similar stories: exploited workers, expropriated land, and a subsidy-hungry agri-business sector eager to divert farmland so that it can be used to feed the world's hunger for motor fuel. Ziegler's outrage reaches a crescendo when he declares that "biofuels are catastrophic for society and the global climate . . . On a planet where a child under age ten dies of hunger every five minutes, to hijack land used to grow food crops and to burn food for fuel constitutes a crime against humanity."[80]

Burning food for fuel is simply a bad idea. But rather than getting into moral discussions, consider just how much food is being used to produce motor fuel. In 2012, about 4.3 billion bushels of US-produced corn—roughly 40 percent of the entire domestic corn crop—was diverted into ethanol production. Bill Lapp, president of Advanced Economic Solutions, an Omaha-based commodity consulting firm, estimates that American motorists are now burning about as much corn in their cars as is fed to all of the country's chickens, turkeys, cattle, pigs, and fish *combined*. Need another comparison? In 2012, the American automobile fleet consumed about twice as much corn as is grown in the entire European Union—put another way, the US ethanol sector is burning about as much corn as is produced by Brazil, Mexico, Argentina, and India *combined*.[81]

To wind down this section, consider Kiefer's conclusion to his report in *Strategic Studies Quarterly*. He wrote that it is time for "leaders and policymakers to catch up with the science and adjust their energy and security strategies to match the objective facts."

The objective facts about biofuels—their low power density, their impact on food prices, their inability to provide even a small fraction of our energy needs—have been known for years. It's well past time for the "greens" to recognize those facts. It's well past time for people who are truly concerned about the environment to recognize that when it comes to energy production, we need density, and the more the better. Biofuel production is not dense. Biofuel production diverts arable land from food production and from nature. Biofuel production is the antithesis of green.

21

CLIMATE CHANGE REQUIRES N2N (N2N IS SFLDC)

Carbon dioxide emissions and climate change have been *the* environmental issues of the last decade or so.

Over that time period, Al Gore became world-renowned for his documentary, *An Inconvenient Truth.* The former US vice president won an Oscar and an Emmy.[1] In 2007, he, along with the Intergovernmental Panel on Climate Change, collected a Nobel Peace Prize for "informing the world of the dangers posed by climate change."[2] That same year, the IPCC released its fourth assessment report, which declared that "most of the observed increase in global average temperatures since the mid-20th century is *very likely* due to the observed increase in anthropogenic GHG [greenhouse gas] concentrations."[3] (Emphasis in original.)

In 2009, Copenhagen became the epicenter of a worldwide media frenzy as some five thousand journalists, along with some one hundred world leaders and scores of celebrities, descended on the Danish capital for a climate change conference that was billed as the best opportunity to impose a global tax or limit on carbon dioxide. What was the result from that meeting, as well as others that were held in Durban, Bonn, and Cancun? Nothing, aside from promises by various countries to get serious—really serious—about carbon emissions sometime soon. In 2011, Pilita Clark, a reporter for the *Financial Times,* summed up the lack of serious action on carbon dioxide emissions, writing that the discussion about global climate issues has "declined into an obscure, jargon-filled process whose impotence was painfully exposed at the 2009 Copenhagen meeting, and which seems even more irrelevant amid today's economic and financial chaos."[4]

Regardless of what you think about carbon dioxide or the climate-change debate, it's apparent that the best way forward is to embrace N2N: natural gas to nuclear.

My position about the science of global climate change is one of resolute agnosticism. I'm not an "alarmist" or "denier." There's no question that carbon dioxide is a greenhouse gas. What we don't know for certain is the ideal concentration of that gas in the atmosphere. I can't talk knowledgeably about polar vortexes, cosmic rays, ice cores, forcings, or aerosols. Nor can I be certain that the climate models being used are accurate. I've become bored by the arguments about "hockey sticks," proper thermometer siting, and whether temperatures have leveled off in recent years.[5] In my view, the media and pundits are way too focused on climate models and not nearly focused enough on reactor, engine, and fuel cell models.

Over the past few years, the discussion about climate change and carbon dioxide emissions has devolved into a hyper-partisan slugfest that's obsessed with tribalism. And that tribalism has obscured nearly everything else. As conservative columnist Mona Charen has explained, "The warmists cast everyone on the other side as paid shills for energy companies, and the skeptics charge that warmists are chasing grant money."[6]

I'm disgusted with the tribalism and the name-calling. I am not interested in being part of anyone's tribe. Put me in a tribe of one: the climate agnostic tribe. I am infinitely more interested in finding and exploiting the innovations that can help us surmount the challenges we face than I am in the endless accusations and name-calling generated by the cadre of self-appointed scorekeepers who spend their entire careers blogging about who might belong on Team Alarmist or Team Denier.

It's time to focus our inquiry on the key question: if we agree that too much carbon dioxide is bad for the Earth's atmosphere—and therefore, for us—what are we going to do? What's the best "no regrets" climate policy as we move forward? As I argued in *Power Hungry,* that policy should be N2N. Adopting N2N makes sense for myriad reasons, but first among them is this: natural gas and nuclear are the fuels of the future because they are so good at producing cheap, abundant, reliable flows of the energy type we crave over nearly every other: electricity.

It's time to put aside the tribalism because, like it or not, the global energy story of today is coal. Unless or until there is a fuel source that can produce electricity at a price that is Cheaper than that now produced from coal, global carbon dioxide emissions will continue to soar. If we want to reduce the rate of growth in carbon dioxide emissions, we must try to make cleaner electricity Cheaper. To do that means embracing the only known lower-carbon fuels that are able to both compete with coal and to make a significant dent in humanity's insatiable appetite for energy and power of all kinds. That means N2N.

Before taking a deeper look at nukes and why climate-change efforts must focus on N2N, I've listed some incontrovertible points about carbon dioxide and climate change, likely wearing out the word "regardless" in the process.

Regardless of whether you think carbon dioxide is causing dangerous fluctuations in the Earth's weather—or that it doesn't matter at all—one fact is clear: we will need vastly more energy in the decades ahead in order to raise the living standards of the more than two billion people who are still living in abject energy poverty.[7]

Regardless of what you think about carbon dioxide emissions, we're going to have to do even more with even less. The prospect of climate change—on a planet that might be getting hotter in some regions, or colder in others, or one that may be having more extreme weather events—is going to accelerate the trend toward Smaller Faster Lighter Denser Cheaper. We are going to need Smaller Lighter Cheaper air conditioners, water pumps, desalination units, computers, fuel cells, turbines, and lighting. We are going to need Smaller Faster Lighter Denser Cheaper engines of all kinds.

Regardless of what you think about carbon dioxide emissions, those emissions are rising, dramatically so. Between 2002 and 2012—the decade of Al Gore—those emissions jumped by 32 percent. That increase reflects soaring global demand for electricity (up by 39 percent), which in turn fostered a 55 percent increase in coal consumption. (Natural gas use increased by 31 percent, while oil use grew by about 14 percent.)[8] Carbon dioxide emissions are growing because people—all seven billion of us on this planet—want electricity. And for many countries, the cheapest way to produce electrons is by burning coal. Carbon dioxide emissions are rising because tens of millions of people around the world are moving out of the dark and into the light, and they're using coal to do it.

Regardless of what you think about carbon dioxide emissions, there are still no affordable, scalable substitutes for the vast quantity of hydrocarbons we use today, and there won't be, not for decades to come. Hydrocarbons now provide about 87 percent of the world's total energy needs.[9] Replacing them would require a new, zero-carbon energy form that can supply about 218 million barrels of oil equivalent per day, an amount of energy equal to the oil output of twenty-six Saudi Arabias.[10] That won't happen because as Vaclav Smil has rightly explained, energy transitions are "deliberate, protracted affairs . . . There is no Moore's law for energy systems."[11]

Global greenhouse gas emissions will continue increasing because global energy demand will likely increase by about 50 percent over the next two decades, and the vast majority of that new demand will be met with hydrocarbons. (By the way, if you think the expected increase of 50 percent by 2035 seems large, recall that between 1900 and 2010, global energy demand increased by about 2,200 percent.)[12]

Regardless of what you think about carbon dioxide emissions, people in the industrialized countries cannot and should not hinder the efforts of the world's poor to gain access to cheap, reliable sources of energy. In other words, rich Greens cannot and should not impede the soaring energy use that's happening everywhere from Bhutan to Borneo. Sure, solar panels and windmills (and geothermal and biomass) are appropriate choices for some locations. Solar, in particular, will grow enormously in the years and decades ahead because it works well in rural and extremely remote areas. All of that said, it's also true that for the vast majority of the world's population, the cheapest and most reliable forms of energy are, and will continue to be, hydrocarbons. According to projections from the Energy Information Administration, while demand for energy over the next twenty-five years or so will be relatively flat in the developed countries (members of the Organisation for Economic Co-operation and Development), the big demand growth

Global Energy Demand Since 1990 and Projected to 2035

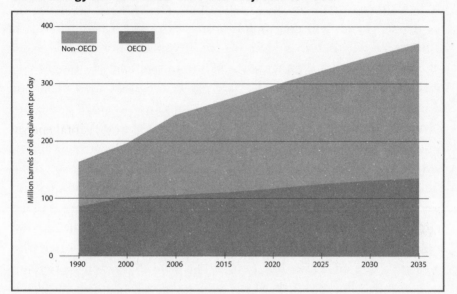

Energy demand in the developed world (countries that belong to the Organization for Economic Cooperation and Development) will rise only slightly over the next twenty-five years or so. The boom in demand—and a concurrent increase in carbon dioxide emissions—will come from developing countries such as China, India, Vietnam, Thailand, Indonesia, and many others. *Source*: Energy Information Administration, International Energy Outlook 2011.

will be in the developing world. By 2035, energy use in the developing countries will likely be around 230 million barrels of oil equivalent per day—nearly double the amount those countries used in 2008 (124 million barrels of oil equivalent per day).[13]

Regardless of what you think about carbon dioxide emissions, we humans must continue adapting and hardening our cities, networks, and structures so they can better survive the always-wacky weather. That means we have to spend money on adaptation. In March 2013, a panel of his science advisers sent President Obama a list of recommendations on climate change. First on that list: focus on national preparedness for climate change.[14]

Don't assume this is a new priority. We humans have been adapting to the climate for millennia. Every sensible "no-regrets" climate policy recognizes the need to prepare for future storms and droughts. We can argue about whether severe weather is caused by carbon dioxide—and whether or not such weather is increasing in frequency or intensity—until the cows come home. The hard reality is that we must make our cities and systems more resilient. Whether those weather events are related to anthropogenic carbon dioxide doesn't matter. What matters is our preparedness with early-warning systems, flood-control measures, and evacuation plans. Few structures can survive an F5 tornado on the Oklahoma plains or the fierce rain and flooding that came with Super Storm Sandy. But we must be prepared so that we can mitigate as much damage as possible and be ready to respond effectively when severe weather strikes, as it surely will. Such adaptation-insurance policies are already paying off. Cities like London, St. Petersburg, and Rotterdam have installed floodgates that help reduce the risk of storm surges. In mid-2013, New York mayor Michael Bloomberg announced a $20 billion adaptation plan for the Big Apple that includes storm-surge barriers and levees.[15]

N2N provides the best no-regrets energy policy because those fuels can provide significant environmental benefits with relatively low economic costs. Natural gas and nuclear are lower-carbon than oil or coal. They emit almost zero air pollution. Better yet, both sources have high power densities, require relatively little land, and can be scaled up enough to meet a significant portion of the continuing growth in electricity demand in places like Hanoi and Seoul.

Is N2N a perfect solution? No. There are no perfect energy sources. All of them come with tradeoffs. But N2N makes the most sense going forward because it can help save money. The table below shows that using natural gas for electricity generation means Cheaper electricity for US consumers.

N2N is positive for air quality. Unlike oil- or coal-fired systems, natural gas–fired engines and turbines emit zero soot and only trace amounts of traditional air pollutants. Natural gas's clean-burning characteristics are helping it steal market share from diesel fuel in the

Estimated Cost of Electricity for Generation Plants Entering Service in the United States in 2018, in US Dollars

Type	Cost per megawatt-hour
Coal	100.1
Natural Gas[16]	65.6
Nuclear	108.4
Geothermal	89.6
Biomass	111.0
Non-dispatchable Technologies	
Wind (Onshore)	86.6
Wind (Offshore)	221.5
Solar Photovoltaic	144.3
Solar Thermal	261.5
Hydroelectric	90.3

When all costs are factored in—transmission, capital, operations and maintenance, etc.— natural gas continues to be the fuel of choice for electricity production in the United States because it is Cheaper than other sources. *Source*: Energy Information Administration, Annual Energy Outlook 2013.[17]

transportation market. Lower emissions and diversification away from oil-fired engines have led to the fuel's adoption in city buses, delivery trucks, and increasingly, long-haul trucks. Iran now has the world's largest fleet of natural gas vehicles, with nearly three million vehicles running on methane. For comparison, the United States has about

120,000 natural gas–fueled vehicles. It ranks seventeenth among the world's countries in that category.[18]

Regardless of what you think about climate change, N2N is helping cut carbon dioxide emissions. When burned in an electric-generation plant, natural gas emits about half as much carbon dioxide as does coal. In May 2012, the International Energy Agency reported that US carbon dioxide emissions had fallen by 92 million tons, or 1.7 percent, since 2011, "primarily due to ongoing switching from coal to natural gas in power generation." The Paris-based agency continued: "US emissions have now fallen by 430 million tons (7.7 percent) since 2006, the largest reduction of all countries or regions." The International Energy Agency credited the shale gale, saying the reduction came from "a substantial shift from coal to gas in the power sector."[19] In other words, market forces in the United States—read: the flood of methane unleashed by the shale gale—have done more to cut carbon dioxide emissions in America than all of the government-mandated programs in Europe.

Other analysts are coming to similar conclusions. In 2013, Max Luke, a policy analyst at the Breakthrough Institute, estimated that the combination of natural gas and nuclear energy has reduced America's carbon dioxide emissions by about 54 billion tons over the last six decades. For comparison, Luke found that wind, solar, and geothermal reduced emissions by just 1.5 billion tons over that same period.[20]

Natural gas–fueled vehicles help reduce greenhouse gas emissions. NGVs emit 20 to 30 percent less carbon dioxide than comparable diesel- and gasoline-fueled vehicles.[21] In mid-2013, the International Energy Agency estimated that global demand for natural gas in the transportation sector will nearly double, to about 9.6 billion cubic feet per day, by 2018.[22] The agency also revealed this remarkable fact: "The expansion of gas as a transport fuel has a bigger impact on reducing the medium-term growth of oil demand than both biofuels and electric cars *combined*."[23] (Emphasis added.)

The biggest increases in natural gas as a transportation fuel are happening in China, where the government is pushing hard to improve air quality in its cities. According to the International Energy Agency, the ramp-up of gas as a transport fuel in China is happening about four times as fast as the growth in the United States.[24]

Cheaper natural gas is gaining market share in the transportation market thanks to an alphabet soup of fuels—NGL, CNG, LNG, DME, GTL. (That's natural gas liquids, compressed natural gas, liquefied natural gas, dimethyl ether, and gas to liquids.) Cheaper natural gas is not only gaining share in the transportation market, it's also becoming more popular at the wellhead to replace the diesel fuel that runs the generators and pumps used on drilling rigs. Numerous drilling companies have begun using natural gas for that purpose, because natural gas, on a per-joule basis, is significantly Cheaper than diesel fuel.

Regardless of what you think about natural gas, the processes used in the drilling, hydraulic fracturing, and transportation of the fuel are getting better. A lot better. They are improving because it is in the interests of the companies that produce oil and natural gas from the earth to make those processes better.

Regardless of what you've heard, hydraulic fracturing is safe. Have there been some cases of groundwater contamination in the Oil Patch? Absolutely. And just about every one of them is due to something other than fracturing. Even the head of the US Environmental Protection Agency, Lisa Jackson, has said the process is safe. In 2011, in testimony before the House Oversight and Government Reform Committee, Jackson said, "I'm not aware of any proven case where the fracking process itself has affected water."[25]

Yes, natural gas is a potent greenhouse gas. And yes, the industry needs to reduce the quantity of gas that leaks from its pipelines and other equipment. But the gas industry in the United States has a big incentive to reduce leakage and therefore have more product to sell. More effort is needed to help reduce methane leakage from other big gas-producers like Russia, where Gazprom's pipes are notoriously leaky.

Regardless of what you think about carbon dioxide emissions or the issue of climate change, it's obvious that N2N fits perfectly with the thesis of this book. In other words, N2N is Smaller Faster Lighter Denser Cheaper.

Natural gas development has a Smaller footprint than wind energy. The areal power density of a marginal natural gas well—one producing 60,000 cubic feet per day—is 28 watts per square meter.[26] That's far higher than the power density of wind and solar energy. It's true that natural gas and oil development requires lots of land. And it's certainly true that there have been many conflicts over the surge in drilling, particularly as drill rigs have begun sprouting in suburban neighborhoods, and in some cases, even inside cities like Fort Worth. But natural gas's land needs are minuscule when compared to the countryside-devouring sprawl that is the hallmark of wind energy.

Natural gas is both Lighter and Denser than other fuels. Because it exists in a gaseous state, natural gas appears Lighter than fuels like wood, coal, or oil. But when measured by weight, gas actually has greater energy density than nearly any other fuel in common use. The gravimetric energy density of natural gas is 53 megajoules per kilogram. For comparison, diesel fuel has 46 megajoules per kilogram.[27]

By now, it should be obvious why natural gas consumption will continue to grow in the years ahead. And to the long list of reasons I've catalogued above, let me add one more: natural gas is the fuel of the future because the Earth contains massive quantities of it.

In November 2009, the International Energy Agency estimated that recoverable global gas resources totaled about 30,000 trillion cubic feet.[28] At current global rates of consumption, that's enough gas to last 250 years.[29] And more shale resources are being discovered all the time. In mid-2013, the British Geological Survey estimated that the Bowland Shale formation in northwest England holds some 1.3 trillion cubic feet of natural gas. The agency's estimate was double the previous estimates.[30]

Natural gas is the fuel of the present. It's also the fuel of the future because gas is not just abundant, it is superabundant. That abundance makes it affordable. And because the methane molecule, CH_4, has only one carbon, it's clean burning. What's not to like?

Let me end the discussion of the first N so we can move on to the second N: nuclear—because if it's not going to be natural gas, it's got to be nuclear.

We Need to Reduce Gas Flaring

Regardless of what you think about carbon dioxide emissions, we need to reduce the flaring of gas. Gas flaring—from Kirkuk and Baghdad to Port Harcourt and Williston—wastes a valuable resource. Globally, according to the World Bank, some 140 billion cubic meters (13.5 billion cubic feet per day) of gas were flared in 2011. That gas is being flared because the producers of the fuel don't have an economic way of getting their product to the market. For many hydrocarbon producers, the dominant economic value of the well's output comes from the oil. And rather than spend the money needed to capture and transport the natural gas that is produced alongside the oil, they simply burn it.

The five biggest wasters of gas: Russia, Nigeria, Iran, Iraq, and the United States. Those five countries account for about 57 percent of the global total. Russia alone is flaring more than 3.6 billion cubic feet of gas per day,[31] an amount of gas that could almost supply all of France's natural gas needs.[32] The amount of gas flared in Iran (1.4 billion cubic feet per day) could nearly supply all of Belgium's gas demand.[33] The amount of gas flared in the United States (about 700,000 cubic feet per day) could nearly supply all of Vietnam's needs.[34] In all, the amount of gas being flared every day around the world is more than enough to supply all of the current natural gas consumption of Africa (about 12 billion cubic feet per day in 2012).[35] In southern Iraq alone, some 700 million cubic feet per day of gas is being flared,[36] even though industry throughout the Middle East is starved for the fuel.

The flaring problem can easily be understood by looking at North Dakota, which has seen a huge boom in oil production from the Bakken Shale. By early 2013, gas flaring in that state was about 300 million cubic feet per day.[37] That's approximately equal to the natural gas consumption of Finland.

On a global basis, we're flaring natural gas that's equal to about 2.5 million barrels of oil per day. That amounts to about 1 percent of all global energy demand.[38] The key innovation, the killer app (along with super-cheap fuel cells, and super-cheap, super-dense electricity storage) is a Cheaper method of turning natural gas into liquid fuel. And that conversion system has to be packaged into a system that can easily be replicated and easily transported to individual well sites where the gas is being flared. Such a gas-to-liquids technology would allow us to convert energy that is wasted into useful product and also reduce carbon dioxide emissions.

22

EMBRACE NUCLEAR GREEN

To call one's self an environmentalist while campaigning against nuclear power (and thus, in a direct and unavoidable way, in favor of coal power) is no longer possible.

—Graham Templeton[1]

In the wake of the 2011 accident at the Fukushima Daiichi plant, it may sound odd to say so, but here goes: the prospects for nuclear energy have never been brighter.

Nuclear has a bright future for several reasons. First among them: as bad as the accident at Fukushima was, the actual damage was pretty well contained. In addition, reactor technology is rapidly improving, the nuclear sector is getting significant private-sector investment, and mainstream environmentalists are embracing nuclear like never before. Furthermore, nuclear energy is Smaller Lighter Denser than all of its competitors. And with the right policies in place, nuclear should get Cheaper.

Nuclear reactors have Smaller footprints because they have very high power densities. The areal power density inside the core of an average reactor is about 338 megawatts (338 million watts) per square meter.[2] The compactness of the design can be seen in the two reactors at the Indian Point Energy Center in Westchester County, New York. Those reactors, with 2,069 megawatts of generation capacity, provide as much as 30 percent of all the electricity needed by New York City.[3] If you include the entire footprint of Indian Point—about 240 acres—the areal power density at the site exceeds 2,130 watts per square meter, meaning that the nuclear plant has 2,100 times as much power density as wind energy (which is 1 watt per square meter).[4]

To equal the electricity generation capacity at Indian Point with wind energy, you'd need to pave about 2,000 square kilometers (772 square miles) with wind turbines, an area three-quarters the size of the state of Rhode Island.[5] Of course, that capacity would still need to be backed up by a natural gas–fired power plant.

Nuclear is superior to other forms of energy production because of its unsurpassed power density. No other form of energy comes within a light-year of nuclear when it comes to the amount of energy it can produce from a small amount of space. The gravimetric energy density of uranium enriched to 3.5 percent and used in a nuclear reactor is roughly 87,000 times that of gasoline.[6] Add in nuclear's minimal carbon-dioxide emissions, and it becomes clear that nuclear can, and will, be providing a significant chunk of the world's electricity for decades—and yes, centuries—to come.

Let me be clear: I'm not claiming that we will see a big surge in new reactor construction in the next few years. Widespread deployment of nuclear energy will take decades. Yes, nuclear is clearly one of our best no-regrets options. But it's also clear that widespread deployment of reactor technology faces huge challenges. The biggest among them: it's still too expensive.

Once they are built, nuclear plants can produce electricity at relatively low cost, but the upfront price tag for the reactors themselves is staggering. In 2012, the US Nuclear Regulatory Commission approved the construction of the Vogtle 3 and 4 reactors, near Augusta, Georgia. The Vogtle reactors, which are primarily owned by Southern Company, will be capable of producing 2,200 megawatts of electricity. The reactors are the first to get a construction permit in the United States since 1978. The reactors will be Westinghouse's AP1000 design, and the total cost of the project is estimated at $14 billion.[7] Thus, building a new nuclear plant in the United States currently costs about $6.3 million per megawatt. For comparison, a coal-fired power plant costs roughly $3 million per megawatt, and a natural gas–fired power plant costs about $1 million.[8]

Investors in the Vogtle reactors estimate that when finished, the nuclear plant will produce power for about eight cents per kilowatt-hour.[9]

And while that's a competitive price, it's readily apparent that major efforts are needed to make the upfront costs of nuclear Cheaper. Another big challenge: the world's biggest environmental groups continue to be nearly unanimous in their opposition to nuclear. They continue their opposition even though nuclear offers the only lower-carbon alternative (aside from natural gas) that can displace significant amounts of coal and do so relatively soon, meaning within a decade or two. The environmental groups' opposition to nuclear proves, once again, that if you are anti–carbon dioxide and antinuclear, you are pro-blackout.

Another essential point: we are just at the beginning of the Nuclear Age. When compared to other power sources, nuclear energy is an infant. And yet, the antinuclear Left wants to kill it in the crib.

Coal has been in use by humans for millennia. It's been in common industrial use for about three hundred years. The history of human use of oil goes back centuries. The adventurer Marco Polo reported seeing oil that was collected from seeps near Baku being used for medicinal purposes as well as for lighting.[10] Petroleum was used to light street lamps in Poland in the 1500s.[11] Natural gas provided lighting for the courthouse in Stockton, California, back in 1854.[12] The history of hydrocarbons makes nuclear look like the toddler it is. The same is equally true for renewables.

Wind? Windmills have been in use for a millenium.[13] Solar? The photovoltaic effect was first observed in 1839. The first solar-photovoltaic device was introduced by Bell Labs in 1954.[14] Biomass? We humans have been burning wood since the discovery of fire some 800,000 years ago.[15] We humans have been relying on renewable energy for thousands of years. And what did we learn in all that time? We found that renewable energy stinks.

Now here comes nuclear energy, a form of electricity production that's only slightly older than I am. And yet, the catastrophists are claiming that nuclear energy is too dangerous and too expensive. They want us to believe the Nuclear Age is over. It's not. It's only just started.

The world's first commercial nuclear plant was Calder Hall, which began producing electricity in Britain in 1956. It produced

October 10, 1956: The reactor vessel arrives at the Shippingport Atomic Power Station in Beaver County, Pennsylvania. The Shippingport facility would become the first commercial nuclear reactor in the United States. The facility, which had an initial capacity of 68 megawatts, began operating in 1957. It successfully operated for more than two decades.[16] *Source*: Library of Congress, HAER PA, 4-SHIP, 1—8.

just 40 megawatts of electricity.[17] A year later, in 1957, the first commercial reactor in the United States began operating at Shippingport, Pennsylvania.[18]

Despite nuclear energy's youth and enormous promise, the antinuclear crowd continues its fear-mongering. On its Web site, Greenpeace makes the outrageous claim that there is "no such thing as a 'safe' dose of radiation."[19] Never mind that we humans are hit with radiation every day of our lives from the sun and from the environment around us. We can count on Greenpeace and the rest of the antinuclear establishment to continue to denigrate nuclear, because fear sells. They claim that nuclear is too dangerous. Debunking that claim only requires a look at the facts about Fukushima.

From a nuclear safety scenario, it's difficult to imagine a scarier scenario than what happened on March 11, 2011. A massive earthquake measuring 9.0 on the Richter scale hit 130 kilometers off the Japanese coastline. Within minutes of the earthquake, a series of seven tsunamis slammed into the Fukushima Daiichi nuclear plant. Some of them were as high as 15 meters. The backup diesel generators, designed to keep the nuclear plant's cooling water pumps operating, quickly failed. A day later, a hydrogen explosion blew the roof off the Unit 1 reactor building. Over the next few days, similar explosions would hit Units 2 and 3.[20] Three reactors melted down.[21]

It was the worst nuclear accident since the Chernobyl accident in 1986. There was widespread fear about the potential for large numbers of casualties due to radiation from the stricken plant. But here's the reality: the accident at the Japanese nuclear plant led directly to exactly two deaths. About three weeks after the tsunami hit the reactor complex, the bodies of two workers were recovered at the plant. They drowned.[22]

For decades, we have been conditioned to believe that radiation is scary. In the wake of the Fukushima accident, there were widespread fears that huge amounts of radioactive materials from the plant would contaminate large areas of Japan and that those same materials could hit the United States. That didn't happen. In early 2013, the World Health Organization reported that radiation exposure due to Fukushima was low. The report concluded: "Outside the geographical areas most affected by radiation, even in locations within Fukushima prefecture, the predicted risks remain low and no observable increases in cancer above natural variation in baseline rates are anticipated."[23]

A few months after the WHO report was published, the UN's Scientific Committee on the Effects of Atomic Radiation released its own report. "No radiation-related deaths have been observed among nearly 25,000 workers involved at the accident site. Given the small number of highly exposed workers, it is unlikely that excess cases of thyroid cancer due to radiation exposure would be detectable in the years to come." The UN committee was made up of eighty scientists from eighteen countries. In addition to finding no documented deaths, the document also praised the actions of the Japanese government immediately after the

2011 accident. "The actions taken by the authorities to protect the public (evacuation and sheltering) significantly reduced the radiation exposures that would have otherwise been received by as much as a factor of 10."[24]

I am not minimizing the seriousness of what happened at Fukushima. The reactors used at the site were of an older, inferior design that lacked the kind of passive-cooling systems that are now being incorporated into reactors. (Passive-cooling systems could have prevented the reactors at Fukushima from melting down.) Furthermore, it's clear that all the problems with the Fukushima reactors have not been solved. In late summer 2013, Tokyo Electric Power Company admitted that it was having difficulty managing more than 200,000 tons of radioactive water being stored in makeshift tanks. Some of those tanks have begun leaking, and some of that leaked water is reaching the ocean.[25] Nor am I forgetting about the huge costs of decommissioning and cleaning up the Fukushima site. In all, the price tag for decommissioning the plant could reach $100 billion, while another $400 billion may be needed to decontaminate areas outside of the plant and to compensate the people who were displaced.[26]

Yes, the price tag for Fukushima will be absurdly high. But this wasn't Chernobyl. The nuclear plants themselves didn't malfunction. Homer Simpson didn't hit the wrong button in the control room. Instead, the reactors at Fukushima Daiichi were hammered by some of the planet's most destructive forces.

The earthquake that hit northeastern Japan on March 11, 2011, was about 700 times as powerful as the killer quake that devastated Haiti in 2010 and left some 300,000 people dead. The Japanese earthquake was the fifth-most powerful one to rock the planet since 1900.[27] The quake was so powerful it affected the rotation of the Earth and shifted the position of the planet's axis by about 17 centimeters (6.5 inches).[28] It's easy to focus on the problems with the nuclear reactors, but the damaged power plants were only a tiny part of the larger devastation. The March 11 earthquake and tsunami killed nearly 16,000 people. It injured another 6,000 or so, and nearly 2,700 people are still missing.[29] Total damages—to infrastructure and the overall Japanese economy— will be measured in the hundreds of billions of dollars.

About ten days after the Fukushima accident, George Monbiot, a veteran environmentalist who had long described himself as "nuclear-neutral," published a column in the *Guardian* to explain that he had changed his mind on the technology. "Atomic energy has just been subjected to one of the harshest of possible tests, and the impact on people and the planet has been small." He continued, "The crisis at Fukushima has converted me to the cause of nuclear power."[30]

While the Fukushima accident has been costly, it has also helped catalyze the push for safer, more resilient reactors. Several companies are already deploying what are known as Generation III+ reactors, which have stronger containment systems and passive safety systems that can cool and stabilize the reactor core for at least three days even if there is no available electricity. Examples of the now-available Generation III+ reactors include the AP1000 from Westinghouse and the European Pressurized Reactor from Areva. What's important here is not necessarily an exhaustive compilation of every known reactor design. Rather, it's to underscore the effort being made to develop reactors that are Smaller Cheaper and safer. Herewith, a short list of some of the most interesting reactor technologies:

Small modular reactors. Generally defined as any reactor with a capacity of 300 megawatts or less, the small modular reactor (SMR) concept is gaining traction for several reasons. First, they cost a fraction of larger reactors. Second, they can be deployed as single or multiple units. If a utility needs, say, 800 megawatts of generation capacity, it could buy as many SMRs as it needs to meet that demand, and the reactors could be added in stages. Third, SMRs are designed to be buried in the ground, which makes them more resistant to natural disasters, terrorism, and mishaps. Finally, and perhaps most important, the SMR could be manufactured in a central location. That final aspect should lead to lower costs, as it would allow the company producing the reactor to maintain a dedicated workforce at one location and ship the reactors—by barge, rail, or truck—to the final destination. Concentrating the workforce in one place should also accelerate the learning curve and allow the company (or companies) producing the reactor to streamline production and reduce costs.

Perhaps the most prominent developer of SMRs is Babcock & Wilcox, which has decades of experience in the nuclear sector. In 2009, the company announced plans to build a modular reactor capable of generating 180 megawatts of electricity. The company's stock is traded on the New York Stock Exchange.

Molten salt reactors. Rather than use fuel rods like conventional reactors, this design mixes the nuclear fuel into a salt mixture. That mix is then pumped in a loop with a reactor on one side and a heat exchanger on the other. When the mixture is in the reactor, it goes into a critical state. The heated salt-fuel mix is then used to produce steam, which, in turn, is used to produce electricity. The design has a fail-safe mechanism in the form of a drain plug at the bottom of the reactor that is made of solid salt. That plug is continually cooled. If the cooler for the plug gets turned off, or if the system's pumps lose power, the plug melts and the molten salt-fuel mix flows into a storage tank where it cools on its own. This design removes the possibility of a meltdown. Molten salt reactors are proven. The Department of Energy tested the design in the 1960s at Oak Ridge National Laboratory, where one ran for six years.[31]

Among the highest-profile promoters of the molten-salt reactors is a start-up company called Transatomic Power, which is promoting what it calls the Waste-Annihilating Molten Salt Reactor.[32] Transatomic is backed by venture capitalists, including Ray Rothrock of Venrock Capital.[33] Transatomic says their reactor design (which exists only on paper) can also run on nuclear waste, a feature that could help deal with the growing volume of spent fuel rods and other nuclear materials being stacked up in locations around the world. (The world now has about 270,000 tons of high-level nuclear waste, and that volume is growing by about 9,000 tons per year.)[34]

Integral fast reactors. A favorite of many nuclear aficionados, the integral fast reactor (IFR) is more than a concept. The Department of Energy built a prototype IFR in Idaho (called the EBR-II) and operated it for three decades. In the 1980s, the agency began building another IFR, but funding for the project was killed by Congress in 1994.[35] The

IFR uses metal cooling instead of water, the coolant used in conventional reactors. The reactor is designed to be safe. Tests on the prototype IFR in Idaho showed that if the reactor's cooling pumps were shut off, the reactor would not overheat. Instead, it simply shut down on its own. In addition, the IFR can burn radioactive waste from other reactors and produce its own fuel. In other words, it can be self-sustaining. The industrial giant General Electric was one of the lead developers of the IFR project with the Department of Energy. Based on its work on the IFR, GE has teamed with the Japanese firm Hitachi to propose what they are calling PRISM, short for Power Reactor Innovative Small Modular. If GE and Hitachi are able to build a PRISM reactor, it would produce about 600 megawatts of power.[36]

Thorium-fueled reactors. Rather than use uranium, some nuclear advocates believe thorium is a superior reactor fuel. Thorium is far more abundant in the Earth's crust than uranium. Unlike uranium, thorium doesn't need to be enriched before it is put into the reactor. Used as a reactor fuel, thorium doesn't produce as many radioactive by-products during fission (such as plutonium) as does uranium. This, in theory, reduces the risk of nuclear proliferation because it cuts the amount of plutonium available for making weapons. In addition, the waste produced by thorium-fueled reactors is far less radioactive than what is produced by conventional reactors.[37] Despite thorium's advantages, however, no commercial operating reactors are using it today.[38] A Virginia-based company, LightBridge, is promoting the use of thorium as a reactor fuel. But the company, whose stock is publicly traded on the NASDAQ, is small—its market capitalization is about $20 million—and is struggling to make money.[39] On the federal side of the ledger, Brookhaven National Laboratory, located on Long Island, New York, has long been a leader in research on thorium as a reactor fuel.

Traveling wave reactors. This design is being pursued by TerraPower, a private company bankrolled, in part, by its chairman, Microsoft founder and billionaire philanthropist Bill Gates, who has put some $35 million into the company. TerraPower's vice chairman is Nathan

Myhrvold.[40] The former chief technology officer at Microsoft, Myhrvold is an author and polymath who has a doctorate in theoretical and mathematical physics from Princeton University.[41] The traveling wave reactor has passive safety features that prevent it from melting down. In addition, it uses sodium as a coolant and depleted uranium (U-238) for fuel. That matters because U-238 is produced as a by-product during the enrichment process for U-235, which is the primary fuel used by conventional reactors. In addition to U-238, TerraPower says their reactor could also be fueled by spent fuel rods from existing conventional reactors, or even thorium.[42] (For the major players in nuclear energy, see Appendix G.)

While private investors and publicly traded companies are seeking opportunities in nuclear, some mainstream environmentalists are finally embracing the technology. Their reason for supporting nuclear is simple: climate change. Their concern about carbon dioxide emissions, along with their understanding that renewable sources like solar and wind cannot begin to provide the scale of energy we demand at prices we can afford, has led them to see nuclear as an essential lower-carbon element of our energy mix.

In 2013, filmmaker Robert Stone released *Pandora's Promise*, a documentary that "explores how and why mankind's most feared and controversial technological discovery is now being embraced by some of the activists who had once led the charge against it."[43] Stone's film is masterly in its use of a simple technique. Stone goes around the world to do interviews, and in many of his stops, including Chernobyl and Fukushima, he carries a handheld Geiger counter that's measuring the background radiation levels. Stone shows that on a beach in Brazil, the background radiation is higher than in some locations near Chernobyl. By doing so, Stone helps demystify radiation and shows that, in fact, there are safe doses of radiation. In fact, we are being hit by radiation nearly all the time. On that beach in Brazil, a family is partially burying an older man in the dark sand because, as the old fellow explains, the radiation in the sand is good for his aches and pains.

Stone's film features environmentalists who had been antinuclear and have since changed their minds, and it devotes significant time to profiling Stewart Brand, the iconoclastic environmentalist who gained fame as the publisher of the *Whole Earth Catalog*, a book that helped define the 1960s and '70s in America. In a trailer for the film, Brand provides a snappy quote: "The question is often asked, 'Can you be an environmentalist and be pro-nuclear?' I would turn that around and say, 'In light of climate change, can you be an environmentalist and not be pro-nuclear?'"

Stone's film also features Michael Shellenberger, the cofounder of the Oakland-based Breakthrough Institute, who is ardently pro-nuclear. The think tank has been a major supporter and proponent of *Pandora's Promise*, and Shellenberger gets the last word in the film, saying that the pro-nuclear forces are gaining strength, and that it feels like "the beginning of a movement."

The Breakthrough Institute has played a key role in catalyzing the pro-nuclear Left. In a 2012 article in *Foreign Policy*, "Out of the Nuclear Closet: Why It's Time for Environmentalists to Stop Worrying and Love the Atom," Shellenberger, along with his Breakthrough Institute cofounder, Ted Nordhaus, and their colleague, Jessica Lovering, summed up the position of the pro-nuclear Greens by declaring that "climate change—and, for that matter, the enormous present-day health risks associated with burning coal, oil, and gas—simply dwarf any legitimate risk associated with the operation of nuclear power plants."[44]

In mid-2013, the think tank published a report that should be required reading for anyone interested in nuclear technology. "How to Make Nuclear Cheap" concludes that making reactors Cheaper will require sustained investment in nuclear technology. That investment should focus on making reactors safer and more modular, meaning that their various components can be standardized and therefore, manufactured at lower cost.[45]

Although the report doesn't single out one reactor design as the "best," it makes a critical point about the need for more governmental involvement. More government commitment is needed to streamline the licensing process for new reactor technologies. It's also needed to enable innovation in materials science. "The history of the commercial nuclear power industry is one in which commercialization in virtually all contexts has depended upon heavy state involvement," states the report. Such governmental involvement is a result of the complexity of nuclear technologies, as well as the need for proper licensing and oversight. Government is also needed to provide insurance in case of a catastrophic accident. "The prospects for accelerating nuclear technology [will] likely depend heavily on the evolving policy and regulatory landscape, both in the United States and abroad."[46]

And therein lies a significant rub. Electricity production from fuels like natural gas and coal doesn't require major interventions from government, because the capital requirements are far lower and the technologies involved are not as complex or potentially dangerous. Jerry Taylor of the Cato Institute has frequently condemned governmental involvement in nuclear, calling nuclear "solar power for conservatives." It's a funny line. But Taylor is ignoring the benefits that nuclear energy can—and should—bring to society.

Our future prosperity depends on cheap abundant reliable supplies of electricity. We should be looking to, and investing in, nuclear because the physics are so favorable. Denser energy is almost always better energy. Nuclear's power-density advantages simply cannot be denied.

Make Atoms for Peace a Reality

In 1946, the Acheson-Lilienthal Report, a document produced for the Truman administration, concluded that the world had entered a new era. "There will no longer be secrets about atomic energy," it said. It declared that given the potential destructiveness of nuclear weapons, there must be "international control of atomic energy" coupled with "a system of inspection."[47] Today, nearly seven decades later, the need for international control of nuclear materials, along with a reliable system for inspecting nuclear-energy plants, remains essential. But the International Atomic Energy Agency remains weak and underfunded.

Created in 1957, the IAEA was designed to be a global "atoms for peace" agency under the aegis of the UN.[48] But it remains a still-obscure entity with no real political muscle. Its 2011 budget was $433.1 million, of which just $171.4 million was spent on nuclear security and nuclear safeguards.[49] Putting that figure into perspective, in 2014, the total US budget for defense-related spending will be approximately $830 billion. That works out to about $2.3 billion per day.[50] Meanwhile, the IAEA, which has responsibility for global nuclear security, is spending just $171.4 million per year, or less than $500,000 per day. In other words, the US military spends 13 times as much *every day* as the IAEA spends on nuclear security *in a year.*

If the United States and the rest of the world are going to embrace nuclear energy as part of a global effort to reduce carbon dioxide emissions, then there must be a concurrent commitment to vigorous international regulation and policing of the nuclear sector. That requires closer monitoring of the fuel used by nuclear reactors as well as increased efforts at ports and other locations to detect radioactive materials that could be used for nefarious purposes.

In 2011, shortly after the meltdown of the reactors at Fukushima, Richard Lester, head of the Nuclear Science and Engineering Department at the Massachusetts Institute of Technology, neatly summed up the issue, writing that "one of the most urgent and important tasks facing the international nuclear community is to affirm, articulate and enforce a set of universal principles of effective nuclear governance."[51]

The only entity that can make that governance effective is a robust IAEA. The IAEA needs more money, political support, and technology. The United States should provide all three. Alas, President Obama has been nearly invisible on nuclear policy for the past few years. In his April 2009 speech in Prague, he said that he wanted to "build a new framework for civil nuclear cooperation, including an international fuel bank, so that countries can access peaceful power without increasing the risks of proliferation."[52] That's an easy speech to make. But making that kind of program into a reality requires having a strong, credible, forceful IAEA. Obama has made little effort toward that end.

The need for a strong IAEA has never been more obvious. Iran wants to harness the atom for electricity production. (And sensibly so. Iran has a young and growing population. Half of the population is under thirty-five, and electricity use more than doubled between 2000 and 2012.)[53] Saudi Arabia, the United Arab Emirates, India, Slovakia, and several other countries also want nuclear and are either building, or planning to build, new reactors. The United States must be a leader in giving the IAEA the authority it needs to make the objectives of the Acheson-Lilienthal report into a reality. No other country has as many nuclear reactors. The United States produces about twice as much electricity from nuclear as France does.[54] The United States has some 7,700 nuclear warheads; only Russia has more, with some 8,500.[55] Most, or better yet, *all*, of those warheads could be converted to peaceful means. Power generation has been one of the best ways of converting the highly enriched uranium used in weapons into something truly useful. (For more on this, see the "Megatons to Megawatts" program that converted Russian warheads into the low-enriched uranium needed for use in American reactors.)[56]

If we are going to try to slow the growth of carbon dioxide emissions, we must have more nuclear energy production, and that must happen on a global basis. The IAEA is the only entity that is credible and has the pedigree to be the global nuclear cop. Supporting the IAEA has been in America's long-term interests since the end of World War II.

It's time to make atoms for peace a reality.

23

SX SMALLER FASTER

WHY THE UNITED STATES WILL DOMINATE
THE SMALLER FASTER FUTURE

It's easy to parody the twenty-, thirty-, and forty-somethings who attend the Interactive portion of Austin's South By Southwest Festival. Paying little heed to those behind, in front, or next to them, some attendees blithely walk along the sidewalk, eyes locked on their iPhones or Droids. Oblivious to their surroundings, the clad-in-black technophiles wander along the sidewalk, narrowly averting collisions with other hipsters, cyclists, and streetlight poles.

It might be even easier to parody a guy like Matt Tran, a twenty-something dressed in a hoodie and jeans who was wandering around the festival with a skateboard in his hand. But Tran, a graduate of Stanford, was no parody. He was in Austin in March 2013 pushing Faster Lighter skateboards.

Tran and two other Stanford grads founded Boosted Boards, a company that claims to be selling the world's lightest electric vehicle. At 12 pounds or so, they might just be right. "This will take you 6 miles in San Francisco," said Tran as he flipped the skateboard over to show me the lithium-ion battery pack and the electronic drive system which was controlled by a small device that Tran held in his hand. "We produce the cover for the battery with a 3-D printer." As a few other people gathered around to take pictures of his board, Tran explained that Boosted Boards had raised money on Kickstarter—about $467,000, nearly five times their goal of $100,000. "It has regenerative braking," said Tran, as he pointed to the board's drive system.

How fast can it go? "We limit it to about 20 miles per hour."

Boosted Boards promises to solve "the last mile." As they explain, the last mile or so of transportation—from public transit to people's homes or workplaces—is the most problematic and costly. If people can carry a device that allows them to get from the bus/train/metro stop to their destination in rapid fashion, they could help solve that last mile challenge. Of course, you can't expect everyone to ride a skateboard, powered by batteries or not. Nevertheless, Tran and his partners at Boosted Boards, Sanjay Dastoor and Matt Ulmen, have done the electric bike one better: they removed the bike.[1]

Tran's impromptu sales pitch in the hall outside the main trade show at South By Southwest was the kind of thing that happens every year at the festival. An event that began as a music confab for slackers and Springsteen-wannabes has expanded so much that the Interactive portion of the festival has become bigger than the music and film events that launched it back in the 1980s. In March 2013, I was among 30,621 people who'd paid to attend the Interactive shindig, and throughout the event, the Austin Convention Center provided a showcase for dozens of entrepreneurs who were on the make, hustling, trying to sell their ideas. And those ideas invariably were about Smaller Faster Lighter Denser Cheaper.

Sure, the conference's trade show had booths occupied by high-profile, multinational companies—Amazon, 3M, and Chevrolet—but there were also scads of motivated entrepreneurs, including people like Jaime Emmanuelli and Jon Miller, the owners and founders of Hive Lighting, a start-up company. Standing in front of their brightly lit booth in the trade show, the two thirty-somethings were pitching their high-output lights, which offer a Smaller Lighter Cheaper alternative to the conventional lights that are used for stage and movie productions. Their key technology: plasma bulbs, which use a radio frequency to excite inert gases and take them into the plasma state. That offers advantages in both efficiency—measured in lumens per watt—and in the amount of heat produced by the lamp. Hive's plasma lights use about half the power of metal-halide lights and about a sixth of what's needed by incandescent lights. That efficiency gain can mean big savings for a movie or TV production that's shooting a scene in a typical house.

March 12, 2013: Jaime Emmanuelli and Jon Miller, the owners of Hive Lighting, at the South By Southwest Interactive festival. Hive manufactures and sells plasma lighting systems that are just as bright as conventional systems but are Cheaper to operate because they require far less electricity. They also produce far less heat. *Source:* Photo by author.

Using Hive's lights, the production crew can simply plug their lights into the wall sockets in the house. If the production were using older, conventional lights, they would likely need to rent and manage a portable generator in order to have enough power to properly light the set. That generator would, in turn, require the use of lots of thick electric cables, connector boxes, and other equipment.

In addition, Hive's plasma bulbs don't produce much heat, a factor that makes them doubly attractive for use on movie and TV sets, which are frequently challenged to keep actors cool as they perform in front of hot lights. "Go ahead, put your hand on it," said Jaime as he pointed to one of Hive's biggest lights. I reluctantly did as instructed. The surface of the lens was warm, but not hot, even though the light itself was blindingly bright.

To be certain, none of the entrepreneurs I met at South By Southwest are curing cancer, solving global warming, or turning back the clock on original sin. I'm not arguing that South By Southwest—or my adopted hometown of Austin—will somehow save the world (although the world does seem to be moving here). Nevertheless, what's happening at South By Southwest Interactive exemplifies the best of America's entrepreneurial spirit. Matt Tran, Jaime Emmanuelli, and Jon Miller are building new businesses. They are inventing new products, risking capital, and just plain hustling. It's because of that hustle, the hustle that's evident in the 3.6 million small businesses in America that have fewer than four employees, that America will endure and prosper.[2]

Put simply, America will dominate the Smaller Faster future because it has a long and illustrious history of entrepreneurship and small business. America has the resources—natural, financial, and intellectual—to foster that entrepreneurial talent. America continues to incubate a spirit of creativity and risk-taking that exists in few other countries. In France, a recent poll found that about 70 percent of people under the age of thirty wanted to get jobs working for the government.[3] In the United States, young people don't want to be bureaucrats. They want to start their own businesses and get rich, just as Jack Dorsey did with Twitter or Nick Woodman did with GoPro.

Let me be clear: I'm not about to stand up and wave Old Glory. I've never believed in American "exceptionalism," whatever that dubious term might mean.[4] It's beyond dispute that America faces many challenges. Furthermore, it's obvious that other countries are exerting more influence, both militarily and culturally. Yes, the United States must make room on the world stage for China. Yes, Africa is growing dramatically and will continue its rapid growth for decades to come. Yes, Brazil dominates Latin America and will continue to do so.

America's crushing debt (about $17 trillion in mid-2013), structural unemployment, increasing economic stratification, decaying infrastructure, and bloated bureaucracy all weaken the country.[5] At the very time that the United States needs a solid political system that can deal with hard choices on spending and taxation, the US Congress has devolved into a hyper-partisan institution that has been gerrymandered into gold-plated, ossified irrelevance. The October 2013 budget impasse that led to a two-week shutdown of the federal government provides the most obvious, and depressing, example of the pettiness and partisanship in Washington. Military bloat and overreach—which reached its apogee with George W. Bush's 2003 invasion of Iraq—has been ruinously expensive. The Second Iraq War, costing more than $800 billion, will be remembered as one of the biggest strategic errors in modern US history.[6]

American politicians continue to appropriate hundreds of billions of dollars per year to the Department of Defense, even though it cannot pass a financial audit.[7] Despite this lack of accountability, the United States continues to spend far more on its military than any other country. In 2011, US military spending was $711 billion. That was more than the military budgets of the next thirteen countries combined: China, Russia, UK, France, Japan, India, Saudi Arabia, Germany, Brazil, Italy, South Korea, Australia, and Canada.[8]

The inability and/or unwillingness of Congress to reign in defense spending is emblematic of a federal government that is dysfunctional in nearly every way. That dysfunction is occurring at the very time that the country desperately needs to be fostering innovation. There are no surefire ways to achieve that, but two areas must be addressed: patent reform and education reform.

For centuries, patents have been viewed as an essential element in the protection of inventors and their intellectual property. Abraham Lincoln famously said that the patent system adds the "fuel of interest to the fire of genius." But over the past few years, it's become clear that the US patent system has become too bloated. In 2011, the Patent Office was granting more than 4,000 patents per week, which is about four times as many as it granted in 1980.[9] The profusion of patents on nearly everything including roses—yes, roses—has led to a litigation-prone system in which far too many companies have forsaken innovation and have instead opted to become "patent trolls," entities that use endless lawsuits (or threats of lawsuits) in order to shake down companies that may be infringing on patents held by the troll.

For example, a Delaware-based company called Innovatio IP Ventures LLC has sent letters to about 8,000 US businesses demanding some $2,500 from each one. Innovatio claims that each of those businesses, including some of America's largest hotel chains, are infringing on the company's patents simply because they are offering Wi-fi access. Patent trolls are now costing American business about $29 billion per year.[10] And without reform, those costs will continue to rob the economy of money that could otherwise be put to better use.

In 2011, Alex Tabarrok, a professor at George Mason University, wrote a short book called *Launching the Innovation Renaissance*, which offered a set of prescriptions for patent reform. Tabarrok contends that the United States has made the process of obtaining patents too easy. Among his suggestions are that patents be awarded for shorter periods of time and that the government should offer more prizes for innovation.[11]

The need for education reform is equally obvious. If the United States is going to continue leading the world, it must get better returns on the enormous amount of money it spends on education. The nation spends 7.3 percent of its gross domestic product educating students from the pre-kindergarten level through college. That's the fifth-highest rate of spending in the world. But the United States ranks only fourteenth in college-completion rates.[12]

Education reform won't happen quickly or easily. Teacher unions—along with all public-sector unions—must be brought to heel. Standardized testing, particularly at the grade-school level, has become a crutch for school administrators, often resulting in students being forced to substitute rote memorization for critical thinking skills. Much of the debate over education has focused on the amount of money being spent rather than the best ways to achieve better results. Education reform will mean accepting the fact that not every high school graduate should be going to college. Yes, the nation needs engineers and doctors. It also needs mechanics and welders, bakers and furniture makers.

For a quick, and depressing, window on the failures of American education, watch the 2010 documentary *Waiting for Superman*. Seeing grade-school children and their parents pin their dreams for the future on a lottery in the hope that they will be allowed to get a proper education is both infuriating and heartbreaking.

But even if we acknowledge America's many challenges and the desperate need for reform—in our politics, patents, and schools—we must also agree that even with these many problems, the United States has numerous advantages over other countries.

People all over the world still want to immigrate to the United States, which remains an economic powerhouse. In 2012, the value of US exports totaled nearly $1.6 trillion. Only China and the EU export more.[13] In fact, when we step back and look at the critical factors—demographics, geography, agriculture, economics and finance, institutions that foster innovation, and availability of low-cost natural resources—the United States could scarcely be better positioned.

The United States is the only developed country on the planet that has a large and growing population—an increasingly important fact for the decades ahead. Europe, Japan, and China all have low birth rates and aging populations. As demographer Joel Kotkin points out, over the next four decades or so, the US population will likely increase by about 100 million people. All of those new residents—many of them immigrants or children of immigrants—will add to the US tax base and provide workers for industry. Further, that immigrant population will have babies. And people who are having babies are hopeful about the future.

Compare that hopefulness with the situation in Russia, a country wracked by corruption and suicide. Russia has the third-highest teenage suicide rate in the world; only Belarus and Kazakhstan have higher rates. Every day in Russia, about five citizens under the age of twenty take their own lives.[14] Russia's teen suicide rate is three times the world average.[15] In 2013, Ian Bremmer, head of the Eurasia Group consulting firm, and Nouriel Roubini, professor of economics at New York University, summed up Russia's problems: "Corruption; ham-fisted authoritarian politics; rising state control of the economy; unfriendly policies toward domestic and foreign investors; and lousy demographics are sapping confidence and strength."[16]

China may be an economic tiger right now, but the country's decades-long enforcement of a one-child-per-family policy will, over the coming decades, wreak economic havoc as the country's supply of workers dwindles. (In late 2013, China's state-run news agency, Xinhua, said the country will relax the policy.) Add in China's lousy environmental record, generational corruption, and lack of natural re-sources (water in particular), and the country's challenges become even more apparent. Japan has long been an industrial powerhouse, but it is facing some truly terrible demographics. It has one of the world's low-est birth rates, and its population of 126 million is expected to drop by about one-third by 2060.[17]

The United States not only has good demographics, it's blessed with terrific geography. Unlike China, Russia, Japan, India, and other coun-tries, we don't have any enemies at, or near, our borders. That means we don't have to fortify our borders. Yes, the United States has increased security along the frontiers with Mexico and Canada, but those moves pale in comparison to the nearly constant tensions between, for in-stance, China and Japan, or India and Pakistan.

American farmers are among the world's most productive. The United States grows nearly 17 percent of all global grain. It's a major producer of nearly every agricultural commodity, from eggs and milk to beef and cotton.[18] America's ability to feed itself means Cheaper food for American consumers. It also gives the nation food security, and thanks to overseas demand for American farm products, significant foreign-exchange revenues.

The United States has a huge advantage over the rest of the world in economics and finance. For decades, the US dollar has been unri-valed as a perceived safe haven for investors. That status has survived the economic crisis of 2008 and the ultra-low interest rates that have prevailed since then. The US dollar may be weaker than it used to be, but it remains the world's most respected currency. Several coun-tries, including Panama, Ecuador, El Salvador, Zimbabwe, and several island nations don't print their own currency. Instead, they trade in greenbacks.[19] Foreigners still view the dollar favorably because no other country has such large, liquid, transparent capital markets. For investors all over the world, the United States continues to be an attractive place to put their money because they have a reasonable assurance that their assets won't arbitrarily be seized by the government. In short, the rule of law counts. And the rule of law—for the most part—is respected in the United States.

In addition, the United States has a huge network of venture capitalists and private financiers who are always looking to put money into the Next Big Thing. In 2012, US venture capital firms invested about $27 billion in various companies.[20] That availability of capital provides entrepreneurs and inventors with the opportunity to continue innovating and, for the best of them, the opportunity to take their new product or service to the marketplace.

Despite its many failings, the United States continues to have the world's best schools. America's land-grant colleges, along with its private and public colleges and universities are incubating innovation. In 2011, according to the National Science Foundation, university spending on research and development totaled a record $65 billion.[21] For comparison, that amount of money is nearly equal to the gross domestic product of Puerto Rico. If America's university-related R&D spending were a country, it would rank in the top ninety of all the countries on the planet in terms of GDP.[22]

As proof of America's innovative nature, compare the number of Nobel Prizes awarded to Americans with the number awarded to non-Americans. In 2012, five Americans were awarded Nobel Prizes, a haul that accounted for half of the Nobels awarded that year.[23] In 2011, thirteen Nobels were awarded, six to Americans.[24]

Between 1901 and 2012, the Nobel Foundation named 862 laureates.[25] Nearly 40 percent of them, 338, were Americans. Over that same period, the United Kingdom came in a distant second, with 119 laureates. Even if you combine the UK's 119 laureates with the Nobels won by Russia, Japan, Germany, and France, (for a total of 332) the United States still comes out ahead. China, despite having a population four times that of the United States, is simply not a factor. By 2012, China had won just eight Nobels.[26] Of course, using a single prize like the Nobel as a gauge doesn't reveal everything about a country's ability to innovate. Nor does such a metric predict the future of innovation. Nevertheless, it's important to note that the United States, with less than 5 percent of the world's population, has been home to nearly 40 percent of all Nobel laureates.[27]

America simply does innovation better than any other country. Schools like Stanford, MIT, Harvard, University of Texas, and

University of Michigan are engines of innovation. No other country has such a large number of schools that generate so many new ideas and products. America's schools are continually churning out students and academics who are thinking of Smaller Faster ways to do medicine, engineering—and yes, like Matt Tran and his colleagues at Boosted Boards—skateboarding. That college/government/private-sector innovation engine cannot easily be replicated anywhere else.

In addition to those factors, the United States has cheap energy. The rest of the world doesn't. Electricity in America is far Cheaper than it is in Europe and other developed countries. In 2012, the average household price of electricity among the twemty-seven members of the European Union was $0.26.[28] In Denmark—the country that wind-energy advocates lionize for its progressive policies—a kilowatt-hour of electricity for residential customers cost a whopping 41 cents. In Germany—by far, Europe's biggest economy, largest electricity consumer, and most important manufacturer—the cost was 35 cents. In Spain, another country that has provided huge subsidies to the renewable-energy sector, it was 29 cents. Meanwhile, stateside, the average residential cost of electricity in 2012 was about 12 cents.[29] The average German resident is paying about three times as much for electricity as the average American, and the average Dane is paying about 3.4 times as much.

Industrialists are noticing. In 2012, Jean-Pierre Clamadieu, the chief executive of Franco-Belgian chemical company Solvay, told the *Financial Times*: "The fact that energy is cheap in the United States, and probably will be for a long time, is changing the game." Further, "Electricity's getting more and more expensive in Europe, and some of the decisions that have been announced regarding nuclear energy production will certainly move the price in the wrong direction. For industry this is really a concern." He went on: "It's very difficult to replace nuclear produced electricity, which costs about €40 per megawatt hour, with a wind turbine put far away in the sea, which costs €200 per megawatt hour."[30]

Electricity is only part of the story. You name it—sun, coal, oil, uranium, natural gas, wind—the United States has loads of each. In 2012, the United States ranked:

- first in natural gas production;
- first in nuclear production;
- first in refined oil product output;
- second in coal production;
- second in electricity production;[31]
- second in refined-product exports;[32]
- third in oil production; and
- fourth in hydro production.[33]

Residential Cost of Electricity in the United States Versus Other Developed Countries in 2012

Country	Cost per kilowatt-hour (in US dollars)
Denmark	$0.41
EU	0.26
France	0.19
Germany	0.35
Ireland	0.26
Italy	0.28
Japan	0.26
Netherlands	0.24
Spain	0.29
Sweden	0.25
Switzerland	0.22
UK	0.20
United States	0.12

Sources: Eurostat; International Energy Agency.

Cheaper: Natural Gas Prices in the United States, Germany, UK, and Japan, 1995–2012

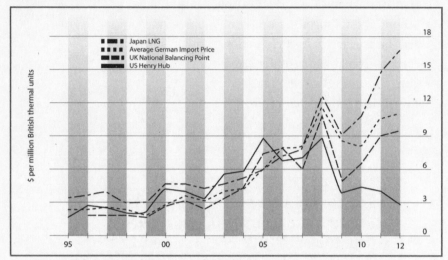

Source: BP Statistical Review of World Energy 2013.

The United States is the world's second-biggest energy producer and second-largest consumer (China ranks first).[34] Lots of people love to hate coal, but the United States has 237 billion tons of coal reserves—about 28 percent of the world's known deposits. That's more than 250 years of supply at current rates of production.[35] The United States isn't the Saudi Arabia of coal, it's the OPEC of coal. America's coal deposits contain 900 billion barrels of oil equivalent.[36] That's nearly as much as the 1.2 trillion barrels of proved oil reserves held by OPEC.[37]

America's energy riches are enormous, and they are increasing thanks to the shale gale, which has fundamentally altered the global industrial balance of power. European and Asian companies are investing tens of billions of dollars in the United States because it has cheap energy. For instance, between 2008 and 2012, foreign companies invested more than $26 billion in the US oil and gas sector.[38]

The shale gale has transformed the US gas sector and turned it from a prospective natural gas importer to an exporter of the fuel. Thanks to the shale gale, the nation now has a price advantage for natural gas that is second to no other country on the planet, with the possible exception of Qatar. Over the past two years or so, US natural gas prices—measured at the Henry Hub in Louisiana—have averaged about $4 or less per million BTU. In the European Union, that same 1 million BTUs of gas will cost three to four times as much. In Japan, it will cost about five times as much.[39]

Cheaper energy has led foreign companies to invest in industrial facilities in the United States. To cite just two examples: an Egyptian company, Orascom, began construction in 2013 on a new $1.8 billion fertilizer plant in Iowa;[40] in June 2013, Vallourec, a French company, opened a new steel mill in Youngstown, Ohio, a Rust Belt town where many of its industrial jobs evaporated over the past few decades. Vallourec's investment in the new steel mill: $1.1 billion.[41]

The United States has cheap gas because it leads the world in production of natural gas—and not by a little bit. In 2012, the United States produced 66 billion cubic feet of gas per day[42]—nearly as much gas as all of the Middle East and all of Africa *combined*. (The 2012 numbers for the Middle East: 52.9 Bcf/d, and Africa: 20.9 Bcf/d.)[43]

What about oil? The United States is leading in that fuel, too. Thanks to the shale gale, US oil production, which has been falling for decades, is suddenly rising, and rapidly. In 2013, production rose by about 1 million barrels per day.[44] The production gains are so dramatic that the United States could soon eclipse both Russia and Saudi Arabia in daily oil production.

America dominates oil and natural gas production because it dominates the global drilling business. That domination has ripple effects throughout the economy, into the railroad, steel, sand, cement, fast food, fuel, trucking, and tire markets, to name just a few. Furthermore, America's dominance will continue for at least a decade to come, and perhaps much longer.

Why? The United States has the rigs, the rednecks, and the pipes.

Shale is the most abundant form of sedimentary rock on the planet.[45] In June 2013, the Energy Information Administration revised its estimates of global shale gas resources. The new estimate: the world has some 23,000 trillion cubic feet of natural gas available in shale deposits. China, Argentina, Algeria, Canada, Mexico, and Australia all have huge shale gas (and shale oil) resources.[46] Those countries, and others, should eventually be able to copy the US model of shale gas production.

But other countries are light-years behind when it comes to developing their own shale deposits, mainly because those countries don't have the rigs needed to develop their shale resources. The United States does.

At any one time, more than half of all the drill rigs on the planet are operating in the United States; in mid-2013, for instance, the US had about 1,750 active rigs. The rest of the world combined had about 1,500.[47] That abundance of rigs allows the United States to drill more wells than any other country. Furthermore, we have the rednecks—and I use that word respectfully—who are able and willing to do the difficult and sometimes dangerous work needed to produce oil and gas. Those men (and some women) are operating and managing those rigs and fracturing operations at all hours of the day and night. Once the rigs and rednecks have finished drilling and fracturing their wells, the hydrocarbons from those wells flow into US pipelines, the world's biggest network with some 1.8 million miles of natural gas pipeline.[48]

Maybe just as important as the rigs, rednecks, and pipes is the issue of mineral rights. America is anomalous in that individuals can own the minerals beneath their feet. No other country allows it. Here, landowners who also own their mineral rights have a huge incentive to encourage drilling on their property as they will get at least one-eighth of the value of the oil and gas produced. In other countries, farmers and rural landowners are naturally opposed to drilling on, or near, their land because they are unlikely to get any economic benefit from the noise, truck traffic, and other disruptions that can come with drilling an oil or gas well.

The punch line of this chapter is this: regardless of which issue you pick—rigs, rednecks, pipes, demographics, geography, agriculture, finance, schools, or mineral rights—America dominates, and no other country even comes close. So don't start saying last rites over the corpse of these United States. Not yet. Yes, America faces plenty of struggles, but it faces even more opportunities. America excels at making things Smaller Faster Lighter Denser Cheaper, and it will continue to excel in the decades ahead. We have the people, the schools, and the resources needed to continue driving the Smaller Faster revolution. We just need to make sure that we use them to our full advantage.

24

CONCLUSION

MOVING PAST FEAR

For years, the message we've been getting from leading environmental-ists has been consistent: reject economic growth, reject modernism, and reject modern forms of energy. And the drumbeat of doom continues. In early 2013, some forty-five years after the publication of his book *The Population Bomb,* Stanford University professor Paul Ehrlich was still pushing his claim that we are facing a catastrophe. In an interview in *Forbes,* Ehrlich was asked what he might say to President Barack Obama if given the chance. Erhlich replied that Obama "should lead the world in showing that economic growth is the disease, not the cure, and shift focus to equity and gross national happiness."[1]

While Ehrlich clings to his neo-Malthusian views, America's most prominent environmentalists continue to demonize natural gas, even though the fuel (along with nuclear) offers the only near-term alter-native to coal. Bill McKibben of 350.org has denounced natural gas, saying that the energy source is not a bridge fuel, it is "just a rickety pier stretching further out into the fossil fuel lake." McKibben also says that hydraulic fracturing, the process used to break up tight geological formations so they can produce oil and gas, must be stopped.[2]

The head of the Sierra Club, Michael Brune, has shown his cata-strophist credentials. Rather than embracing natural gas as a way toward cleaner air and lower carbon dioxide emissions, he's begun calling nat-ural gas an "extreme fossil fuel." Brune claims that gas is "a gangplank to a destabilized climate and an impoverished economy."[3]

While McKibben and the Sierra Club are attacking natural gas, other activists are continuing their protests against genetically modified plants. In May 2013, organizers held the March Against Monsanto, an effort aimed at calling attention to the seed giant's push to produce more genetically modified organisms. Marches were reportedly held in more than four hundred cities. The marches were organized by a Utah-based woman, Tami Monroe Canal, who said the group's protests will "continue until Monsanto complies with consumer demand. They are poisoning our children, poisoning our planet."[4] Shortly before the protest was held, in an interview posted on YouTube, Canal claimed that the company has "no intention of serving the people. They betray humanity. They betray life. They malign Mother Nature. And they do so at the expense of all of us."[5]

What binds the views of Ehrlich, McKibben, Brune, Canal, and many others is their never-ending worry. Their outlook rejects innovation and modern forms of energy. It rejects business and capitalism. Whether the message is explicit or implicit, the message coming from many of the "greens" is an anticorporate, anticapitalist stance that is rooted in the notion that any large business is one to be feared. Monsanto is bad. Halliburton, a company that does hydraulic fracturing, is bad. Meanwhile, in 2013, Apple was named the world's most-admired company by *Fortune* magazine. Never mind that in mid-2013, Apple's market capitalization of about $400 billion was roughly eight times that of Monsanto's and ten times that of Halliburton's.[6]

We have to discard the notion that profits are bad. We must accept—and better yet, celebrate—the tinkerers and entrepreneurs who are designing better batteries, more efficient lights, and better drill rigs in the hope of making money. We must move past the climate of fear to one of optimism. We must move past fear of technology to an understanding that technology isn't the problem; it's the solution.

There's no question that we humans have changed the face of the planet. And we will continue changing it. Some species will disappear, and many of those extinctions will be due to human blundering. Parts of paradise will be paved. But we cannot forsake human creativity, economic productivity, and our efforts to end hunger and poverty, in the hope that doing so will magically restore the planet's ecological systems. We've dammed rivers, plowed the prairies, and drilled oil wells in the rain forests. Activities like that will continue, and as they do, we will see more tradeoffs. But we cannot simply stop economic development. The notion of degrowth being pushed by groups like Worldwatch Institute, Greenpeace, and others, is an affront to human ingenuity and aspiration. The claim that we can all rely on renewable energy—an assertion that we keep hearing from activists like Bill McKibben and biofuel hucksters like Amory Lovins—is a damnable lie.

We humans are not going to retreat back to the grinding poverty that stunted so many lives in the past and continues to do so in the world's least-developed places. We are, most of us, city dwellers. We are not going to "return to the land." Not willingly.

We call our species *homo sapiens*—wise man—but we are, in fact, *homo faber*, man the creator. We have changed the face of this planet with our tools and structures, and we will continue doing so. Assuring future prosperity requires that we continue exploring the atom and exploring deep space. We must continue creating new tools and nourishing new ideas. Our future depends on innovation. We need innovation so we can make energy Cheaper. Cheaper energy will foster better living standards and more innovation. If we are going to reduce the rate of growth in carbon dioxide emissions, we have to disseminate technologies for natural gas production to the countries that lack them. We have to get good at nuclear. And we will. The United States and other countries must continue investing in making nuclear reactors safer and Cheaper so that the power of the atom can be more effectively harnessed.

Smaller computing devices, Faster communications, Lighter vehicles, Denser cities, and Cheaper energy will help foster innovation. Innovation, in turn, will generate the wealth and new ideas and technologies we need to deal with the challenges we face, whether it's a changing climate, food and water scarcity, or difficult diseases. The best way to protect the environment is to get richer. Wealthy countries can afford to protect the environment. Poor ones generally can't.

We need more wealth, more economic growth, more jobs. We need, as Ted Nordhaus and Michael Shellenberger have written, "a new, creative, and life-affirming worldview . . . Wealth and technology liberated us from hunger, deprivation, and insecurity; now they must be considered essential to overcoming ecological risks."[7] British journalist and environmental activist Mark Lynas gives a similar view for why we need such a positive outlook. In his 2011 book, *The God Species*, Lynas wrote, "Only optimism can give us the motivation and passion we will need to succeed. Voices of doom may be persuasive, but theirs is a counsel of despair."[8]

For decades—even for centuries—we've been deadened by the drumbeat of despair. It's time to dismiss the Jeremiahs who are claiming that our redemption lies in rejecting modernity and economic growth. It's time to reject the dystopians, catastrophists, fearmongers, and doomsayers. It's time for an anti-neo-Malthusian outlook. It's time for an outlook that embraces humanism, optimism, technology, and a belief that things are getting better. Such an outlook is not only life-affirming, it also has the virtue of being true. Technology and economic growth have brought—and are bringing—tens of millions of people out of the dark and into the electric-lit world of ideas, education, and fuller, healthier, freer, more fulfilling lives. That same growth is helping to protect nature.

I am—as the late author Molly Ivins used to say—"optimistic to the point of idiocy." And that nearly idiotic optimism springs from the inexorable human desire for Smaller Faster Lighter Denser Cheaper.

SI NUMERICAL DESIGNATIONS

We use many numerical designations—milli, mega, nano—on a regular basis without recognizing that they are part of the International System of Units. The system is commonly known as SI, the abbreviation for Système International d'Unités. (France was instrumental in the effort to harmonize units of measure.) Given that most people are only passingly familiar with these designations, see below for a review of all of the numerical designations—from yocto and yotta.

The difference between yocto and yotta is the difference between a septillionth and a septillion. Between yocto, the SI prefix for 10^{-24}, and yotta (sometimes spelled yota), the SI prefix for 10^{24}, there are 48 zeroes. It's the difference between 0.000,000,000,000,000,000,000,001 and 1,000,000,000,000,000,000,000,000. But in SI, those numbers would be written without the commas, thus, yocto is: 0.000 000 000 000 000 000 000 001; and yotta is: 1 000 000 000 000 000 000 000 000.

Herewith, all of the SI numerical designations and their symbols.

Number	Prefix	Symbol
10^{-24}	yocto–	y
10^{-21}	zepto–	z
10^{-18}	atto–	a
10^{-15}	femto–	f
10^{-12}	pico–	p
10^{-9}	nano–	n
10^{-6}	micro–	μ
10^{-3}	milli–	m

(continues)

Number	Prefix	Symbol
10^{-2}	centi-	c
10^{-1}	deci-	d
10^{1}	deka-	da
10^{2}	hecto-	h
10^{3}	kilo-	k
10^{6}	mega-	M
10^{9}	giga-	G
10^{12}	tera-	T
10^{15}	peta-	P
10^{18}	exa-	E
10^{21}	zetta-	Z
10^{24}	yotta-	Y

Source: http://www.math.com/tables/general/numnotation.htm.

ENERGY AND POWER: UNITS AND EQUIVALENTS

POWER UNITS AND EQUIVALENCIES

1 watt (W) = 0.00134 horsepower or 1 joule/second (J/s)

1 kilowatt = 1,000 watts or 1.35 horsepower (hp)

1 megawatt = 1,000 kilowatts, or 1,000,000 watts

1 gigawatt = 1,000 megawatts, 1,000,000 kilowatts, or 1,000,000,000 watts

1 terawatt = 1,000 gigawatts, 1,000,000 megawatts, 1,000,000,000 kilowatts, or 1,000,000,000,000 watts

1 electric lamp of 100 W = 0.1 kW

1 car engine with a 60 hp engine = 44 kW

1 wind turbine rated at 1 megawatt (MW) = 1,350 hp

1 nuclear plant with 1,000 MW of capacity = 1,350,000 hp

1 gallon of oil equivalent per day = 0.71 hp (529 W)

1 barrel of oil equivalent per day = 30 hp (22.1 kW)

1,000 cubic feet of natural gas per day = 5 hp (3,819 W)

1 day of Saudi Arabia's oil production = 250,000,000 hp (186.5 billion W or 186.5 gigawatts)

ENERGY UNITS AND EQUIVALENCIES

0.1 joule = energy used in average golf putt

1 Btu = energy released by burning 1 wooden match = 1.055 kilojoules

1 cubic foot of natural gas = 1,031 Btu = 1.09 megajoules

1 cubic foot = volume of a regulation basketball

1 cubic meter = 35.3 cubic feet

1 kilowatt-hour (kWh) of electricity = 3,412 Btu = 3.6 megajoules

1 gallon gasoline = 125,000 Btu = 125 megajoules

1 gallon gasoline = 36 kWh of electricity

1 ton of oil = 7.33 barrels (bbl) of oil

1 bbl of oil = 42 gallons or 159 liters

1 bbl of oil equivalent = 5,800,000 Btu = 5.8 gigajoules

1 bbl of oil equivalent = 1.64 megawatt-hours (MWh) of electricity

1 bbl of oil equivalent = 5,487 cubic feet of natural gas

1 kilowatt-hour (kWh) = 1,000 watts for 1 hour

1 megawatt-hour (MWh) = 1,000,000 watts for 1 hour

GRAVIMETRIC POWER DENSITY FROM HUMANS TO JET ENGINES

YEAR	ENGINE/ INVENTOR	PRIME MOVER	WEIGHT (KG)	HORSE- POWER	WATTS	GRAVIMETRIC POWER DENSITY (W/KG)
Prehistory to present	Human[1]		77	0.134	100	*1.3*
Prehistory to present	Horse[2]		450	1	746	*1.7*
Early 1800s	Boulton & Watt[3]	Low-pressure steam engine	1,818	24	17,904	*9.8*
1816	French–Shreve[4]	High-pressure steam engine	4,536	100	74,600	*16.4*
1851	Henry Burden[5]	Utica waterwheel	226,796	300	223,800	*1.0*
1865	Railroad locomotive[6]	Steam engine	27,272	520	387,920	*14.2*
1876	Corliss[7]	Centennial steam engine	50,802	1,400	1,044,400	*20.6*
1886	Karl F. Benz[8]	Internal combustion engine	96	1	500	*5.2*
1903	Wright Brothers[9]	Wright Flyer engine	77	12	8,952	*116.3*
1908	Ford[10]	Model T engine	136	22	16,412	*120.7*
1943	Wright R-3350[11]	Radial engine for B-29 bomber	1,212	2,200	1,641,200	*1354.1*
2012	Ford[12]	EcoBoost	97	123	91,758	*946.0*
2013	Formula 1[13]	Racing engine	95	750	559,500	*5889.5*
2013	GE[14]	GEnx-1B jet engine	5,828	117,000	87,282,000	*14976.3*

FIVE LEADERS IN ONLINE LEARNING

Dozens of companies have sprung up in recent years to promote online learning. This list provides only a summary of some of the more prominent ones.

OFFICIAL NAME	COURSERA
Web site	https://www.coursera.org/
Headquarters	Mountain View, CA
Founded	2011 by two Stanford computer science professors, Andrew Ng and Daphne Koller
Profit Model	For Profit
Ownership	Private
Backer(s) & Revenue	$16 million in venture capital backing, awarded April 2012, by Kleiner Perkins Caufield & Byers
Students	3.2 million from more than 200 countries
University Partners	62
Total Courses	327
Format	Free
	Video presentations
	Short quizzes and testing
Notes	Largest MOOC platform in the world Based on Stanford's original online class platform Coursera leaves the design of the courses up to the individual institutions within broad guidelines

OFFICIAL NAME	edX
Web site	https://www.edx.org/
Headquarters	Cambridge, MA
Founded	2012 venture between MIT and Harvard
Profit Model	Nonprofit
Ownership	Nonprofit
Backer(s) & Revenue	$60 million (MIT and Harvard each contributed $30 million)
Students	370,000, primarily college-age students
University Partners	12
Total Courses	62
Course Format	Free
	Lectures from 3 to 15 minutes
	Quizzes, homework assignments, and labs throughout
	Students must complete the work on deadline, and because everything is auto-graded, feedback is instantaneous
	Classes run 6–12 weeks
	Students and professors connect through discussion forums and chat rooms
Notes	Based on MIT's original online class platform, MITx

OFFICIAL NAME	KHAN ACADEMY
Web site	http://www.khanacademy.org/
Headquarters	Mountain View, CA
Founded	2006 by Salman Khan (MIT and Harvard alum)
Profit Model	Nonprofit
Ownership	Nonprofit
Backer(s) & Revenue	Khan lived off of his savings for the first 9 months, until he received his first significant donation from Ann Doerr.
	Large Initial Donors:
	John & Ann Doerr ($109,000)
	Google ($2 million)
	Bill and Melinda Gates Foundation ($1.5 million)
Students	40 million in United States
	Videos watched: 250 million
	Exercises completed: 1 billion
University Partners	N/A
Total Courses	More than 2,400 videos
Course Format	Free
	Math, biology, chemistry, physics, finance, and history videos
	Each video is approximately 10 minutes long.

(continues)

OFFICIAL NAME	KHAN ACADEMY
Notes	One of the largest educational resources in the world Teaming with Bank of America to offer financial literacy courses for adults

OFFICIAL NAME	TEACHING COMPANY (Product: The Great Courses)
Web site	http://www.thegreatcourses.com
Headquarters	Chantilly, VA
Founded	1990 by Thomas M. Rollins, former chief counsel of the United States Senate Committee on Labor and Human Resources
Profit Model	For Profit
Ownership	Private
Backer(s) & Revenue	By 2011, it had $110 million per year in sales. Sold to Brentwood Associates in 2006 for undisclosed amount
Students	Hundreds of thousands
University Partners	N/A
Total Courses	About 400 courses across 10 subject areas taught by top professors
Course Format	DVD, audio CD, and MPEG-4 downloads and streaming, plus transcript books
Notes	Not a MOOC but an interesting example of education decentralization

OFFICIAL NAME	UDACITY
Web site	http://www.udacity.com/
Headquarters	Palo Alto, CA
Founded	2011 after a Stanford University experiment in which cofounders Sebastian Thrun and Peter Norvig offered their "Introduction to Artificial Intelligence" course online to anyone for free. Cofounder Mike Sokolsky, a robotics researcher, also joined the venture.
Profit Model	For Profit
Ownership	Private
Backer(s) & Revenue	Investors: Charles River Ventures ($5 million) Andreessen Horowitz ($15 million)
Students	750,000
University Partners	N/A
Total Courses	28
Course Format	Free Interactive courses with activities, quizzes, and exercises interspersed between short videos and talks by instructors and industry experts

WIND ENERGY'S NOISE PROBLEM:
A REVIEW

The fundamental problem with wind energy is its low power density. Trying to extract significant quantities of energy from the wind requires large turbines to have blades that move at high speeds. And those blades create low-frequency noise and infrasound that is affecting people living near wind projects. Herewith, a partial summary of recent global developments on the wind-turbine noise problem.

- In 2009, a standing committee of the parliament in New South Wales, Australia, recommended "a two-kilometer minimum setback between wind turbines and neighboring houses."[1] The committee concluded that "reputable research has shown that noise annoyance is an adverse health effect that can result from wind farms, as it can result in effects such as negative emotions and sleep disturbance."[2]
- In 2010, a book published in New Zealand, *Sound, Noise, Flicker and the Human Perception of Wind Farm Activity,* which includes twenty-three peer-reviewed articles by acousticians and engineers, concludes that "the latest research indicates that nuisance noise from wind farms is associated with psychological distress, stress, difficulties with falling asleep, and sleep interruption . . . It is clear that there must be far more care in the siting of any future wind farms and a better understanding of how to mitigate the noise and compensate the affected individuals."[3]

- In September 2010, the *Copenhagen Post* reported that "state-owned energy firm DONG Energy has given up building more wind turbines on Danish land, following protests from residents complaining about the noise the turbines make." The article goes on to quote the Danish wind giant's CEO, Anders Eldrup: "It is very difficult to get the public's acceptance if the turbines are built close to residential buildings, and therefore we are now looking at maritime options."[4]

- In mid-2011, the state government of Victoria, in southeastern Australia, announced that it would enforce a 2-kilometer set-back between wind turbines and homes. The state's planning minister said the setback was needed for health reasons.[5]

- In August 2011, in a peer-reviewed article in the *Bulletin of Science, Technology & Society,* Carl V. Phillips, a Harvard-trained PhD, concludes that there is "overwhelming evidence that wind turbines cause serious health problems in nearby residents, usually stress-disorder type diseases, at a nontrivial rate."[6]

- In October 2011, Alec Salt, an otolaryngology professor at Washington University in St. Louis who has studied the effects of low-frequency noise, says that if industrial wind turbines are placed within a mile of residential structures, "you're asking for trouble."[7]

- In 2011, Ontario's Environmental Review Tribunal conducted an inquiry into a proposed wind-energy facility known as the Kent Breeze Wind Farm Project.[8] Although the officials allowed the facility to be built, the tribunal concluded: "This case has successfully shown that the debate should not be simplified to one about whether wind turbines can cause harm to humans. The evidence presented to the Tribunal demonstrates that they can, if facilities are placed too close to residents. The debate has now evolved to one of degree."[9]

- In 2012, Peter Narins, a distinguished professor and expert on auditory physiology at the University of California–Los Angeles, published a paper in the journal *Acoustics Today*. Narins and his student coauthor, Annie Chen, found that wind turbines generate "substantial levels of infrasound and low frequency sound," and therefore, "modifications and regulations to wind farm engineering plans and geographical placements are necessary to minimize community exposure and potential human health risks."[10]

- In 2012, a peer-reviewed study published in the journal *Noise & Health* found a relationship between wind farms and "important clinical indicators of health including sleep quality, daytime sleepiness and mental health." The epidemiological study compared two groups of Maine residents with similar demographics. The residents who lived near the wind project suffered more interrupted sleep. The study, which was led by a Maine-based radiologist, Michael Nissenbaum, along with two coauthors, also found a "significant" link—probably caused by poor-quality sleep—between wind turbines and poorer mental health. The report's conclusion:

We conclude that the noise emissions of IWTs [industrial wind turbines] disturbed the sleep and caused daytime sleepiness and impaired mental health in residents living within 1.4 km of the two IWT installations studied. Industrial wind turbine noise is a further source of environmental noise, with the potential to harm human health. Current regulations seem to be insufficient to adequately protect the human population living close to IWTs. Our research suggests that adverse effects are observed at distances even beyond 1 km.[11]

APPENDIX F

AREAL POWER DENSITY FOR SIXTEEN WIND-ENERGY PROJECTS

1. ROSCOE WIND PROJECT, ROSCOE, TEXAS.[1]

781.5MW on 100,000 acres.

Calculation: 781,500,000 watts on 400,000,000 square meters = 1.95 W/m^2

2. WAUBRA WIND FARM, NEAR BALLARAT, VICTORIA, AUSTRALIA.[2]

192 MW on 173 square kilometers.

Calculation: 192,000,000 watts on 173,000,000 square meters = 1.1 W/m^2

3. LANGFORD WIND FARM, NEAR SAN ANGELO, TEXAS.[3]

150 MW on 35,000 acres.

Calculation: 150,000,000 watts on 141,640,000 square meters = 1.06 W/m^2

4. LOS VIENTOS WIND PROJECT, WILLACY COUNTY, TEXAS.[4]

200 MW on 30,000 acres.

Calculation: 200,000,000 watts on 121,400,000 square meters = 1.65 W/m^2

5. FLAT RIDGE 2 WIND PROJECT NEAR WICHITA, KANSAS.[5]

470,000,000 MW on 66,000 acres.

Calculation: 470,000,000 watts on 267,000,000 square meters = 1.76 W/m^2

6. FLAT RIDGE 1, NEAR WICHITA, KANSAS.[6]

50 MW on 5,000 acres.

Calculation: 50,000,000 watts on 20,200,000 million square meters = 2.47 W/m^2

7. CHOKECHERRY AND SIERRA MADRE WIND PROJECT, WYOMING.[7]

2,000 to 3,000 MW on 229,000 acres.

Calculation: assume 2,500,000,000 watts on 926,730,000 square meters = 3.2 W/m^2

8. CAPITAL WIND FARM, NEAR CANBERRA, NEW SOUTH WALES, AUSTRALIA.[8]
141 MW on 35 square kilometers.
141,000,000 watts on 35,000,000 square meters = 4 W/m^2

9. SNOWTOWN WIND FARM, NEAR ADELAIDE, SOUTH AUSTRALIA, AUSTRALIA.[9]
99 MW on 12,000 hectares.
Calculation: 99,000,000 watts on 120,000,000 square meters = 0.825 W/m^2

10. RIPLEY WIND POWER PROJECT, NEAR RIPLEY, ONTARIO, CANADA.[10]
76 MW on 3,600 hectares.
Calculation: 76,000,000 watts on 36,000,000 square meters = 2.1 W/m^2

11. ERIE SHORES WIND FARM, PORT BURWELL, ONTARIO, CANADA.[11]
99 MW on 5260 hectares.
Calculation: 99,000,000 watts on 52,600,000 square meters = 1.88 W/m^2

12. GREENWICH WIND FARM, THUNDER BAY DISTRICT, ONTARIO, CANADA.[12]
99 MW on 10,000 acres.
Calculation: 99,000,000 watts on 40,468,000 square meters = 2.44 W/m^2

13. KINGSBRIDGE I WIND POWER PROJECT, GODERICH, ONTARIO, CANADA.[13]
40 MW on 1,000 hectares.
Calculation: 40,000,000 watts on 10,000,000 square meters = 4 W/m^2

14. MELANCTHON I WIND PLANT, MELANCTHON, ONTARIO, CANADA.[14]
67.5 MW on 2,500 hectares.
Calculation: 67,500,000 watts on 25,000,000 square meters = 2.7 W/m^2

15. BP WIND ENERGY: SHERBINO 2 WIND PROJECT, PECOS COUNTY, TEXAS.
150 MW on 20,000 acres.
Calculation: 150,000,000 watts on 80,900,000 square meters = 1.8 W/m^2

16. MEHOOPANY WIND FARM, WYOMING CITY, PA.[15]
144 MW on 9,000 acres.
Calculation: 144,000,000 watts on 36,421,000 square meters = 3.9 W/m^2

Average power density for the sixteen projects is 2.3 watts per square meter. Multiplying 2.3 watts per square meter by 0.39 to account for capacity factor gives an average power density of 0.9 watts per square meter.

MAJOR PLAYERS IN NUCLEAR ENERGY

Note: The data presented here is the most recent available (as of October 2013). In a few instances, the numbers are approximate due to discrepancies in some of the nuclear databases.

OFFICIAL NAME	AREVA USA
Web site	www.areva.com
Ownership	Public (Ticker: AREVA)
Headquarters	Courbevoie, Paris (France)
Market Capitalization	$4.7 billion
2011 Revenue	$11.66 billion
Key Nuclear Technology	European Pressurized Reactor (EPR)
Reactors Built	Operating globally: 100
Reactors Under Construction	4

OFFICIAL NAME	THE BABCOCK & WILCOX COMPANY (B&W)
Web site	www.babcock.com
Ownership	Public (Ticker: BWC)
Headquarters	Charlotte, NC (US)
Market Capitalization	$3.5 billion
2011 Revenue	$2.95 billion
Key Nuclear Technology	Modular Light Water Reactor
Reactors Built	10
Reactors Under Construction	n/a

OFFICIAL NAME	CHINA NATIONAL NUCLEAR CORPORATION (CNNC)
Web site	www.cnnc.com.cn/tabid/141/Default.aspx (English version)
Ownership	State-owned
Headquarters	Beijing, China
Market Capitalization	No financial data available
2011 Revenue	n/a
Key Nuclear Technology	Pressurized Water Reactor (PWR)
	Heavy Water Reactor (HWR)
Reactors Built	6
Reactors Under Construction	8

OFFICIAL NAME	GENERAL ELECTRIC COMPANY (GE)
Web site	www.ge.com www.ge-energy.com (energy division)
Ownership	Public (Ticker: GE)
Headquarters	GE: Fairfield, CT (US)
	GE-Energy: Atlanta, GA (US)
Market Capitalization	$235 billion
2011 Revenue	$147 billion
Key Nuclear Technology	Advanced Boiling Water Reactor (ABWR) Economic Simplified Boiling Water Reactor (ESBWR) Power Reactor Innovative Small Modular (PRISM)
Reactors Built	About 50
Reactors Under Construction	2

OFFICIAL NAME	KOREA ELECTRIC POWER CORPORATION (KEPCO)
Web site	www.kepco.co.kr/eng (English version)
Ownership	Public (Ticker: KEP)
Headquarters	Seoul, South Korea
Market Capitalization	$17.7 billion
2011 Revenue	$43.53 billion
Key Nuclear Technology	APR1400
Reactors Built	14
Reactors Under Construction	8

OFFICIAL NAME	THE NUCLEAR POWER CORPORATION OF INDIA LTD. (NPCIL)
Web site	www.npcil.nic.in (English version)
Ownership	State-owned
Headquarters	Mumbai, India

(continues)

OFFICIAL NAME	THE NUCLEAR POWER CORPORATION OF INDIA LTD. (NPCIL)
Market Capitalization	n/a
2011 Revenue	$1.2 billion
Key Nuclear Technology	Boiling Water Reactor (BWR) Pressurized Heavy Water Reactor (PHWR)
Reactors Built	21
Reactors Under Construction	6

OFFICIAL NAME	STATE ATOMIC ENERGY CORPORATION (ROSATOM)
Web site	www.rosatom.ru/en (English version)
Ownership	State-owned
Headquarters	Moscow, Russia
Market Capitalization	n/a
2011 Revenue	$15 billion
Key Nuclear Technology	VVER Pressurized Water Reactor (PWR)
Reactors Built	33
Reactors Under Construction	10

OFFICIAL NAME	UNITED STATES NAVY
Web site	www.navy.com
Ownership	US Government
Headquarters	Washington, DC (US)
Market Capitalization	n/a
2012 Budget	$173 billion
Key Nuclear Technology	Pressurized Water Reactor
Reactors Built	Through 2010, US Navy had built 219 nuclear-powered ships
Reactors Under Construction	In mid-2013, six more nuclear-powered ships were being built.

OFFICIAL NAME	WESTINGHOUSE ELECTRIC COMPANY
Web site	www.westinghouse.com www.westinghousenuclear.com
Ownership	Private
Headquarters	Cranberry Township, Butler County, PA (US)
Market Capitalization	No financial data available
2011 Revenue	$5.2 billion
Key Nuclear Technology	AP1000 (PWR)
Reactors Built	About 70
Reactors Under Construction	About 17

NOTES

Notes to Introduction

1. Russ Mitchell, "World population will reach 7 billion," CBS News, October 29, 2011, http://www.cbsnews.com/8301–18563_162–20127508/world-population-will -reach-7-billion/. UN News Service, "As world passes 7 billion milestone, UN urges action to meet key challenges," October 31, 2011, http://www.un.org/apps/news /story.asp?NewsID=40257.

2. Erica Bulman, "WHO: Infectious Diseases Spread Faster," Associated Press, August 22, 2007, http://www.washingtonpost.com/wp-dyn/content/article/2007 /08/22/AR2007082202248_pf.html.

3. Maria Cheng, "2 new diseases could both spark global outbreaks," Associated Press, May 13, 2013, http://news.yahoo.com/2-diseases-could-both-spark-global -outbreaks-133527593.html.

4. For more, see terror-alert.com.

5. Gregg Easterbrook, *The Progress Paradox: How Life Gets Better While People Feel Worse* (New York: Random House, 2003), xviii.

6. For an example of this rhetoric see President Obama's June 25, 2013 speech at Georgetown University on climate change: http://www.bloomberg.com/news /2013–06–25/-we-need-to-act-transcript-of-obama-s-climate-change-speech .html. While Obama and many Democrats and environmental groups continue to claim that carbon dioxide emissions are causing more extreme weather events, the latest IPCC report says that is not so. One quote from the report: "In summary, the current assessment concludes that there is not enough evidence at present to suggest more than low confidence in a global-scale observed trend in drought or dryness (lack of rainfall) since the middle of the 20th century due to lack of direct observations, geographical inconsistencies in the trends, and dependencies of inferred trends on the index choice." For more on this, see Roger Pielke Jr.'s work, and in particular this blog entry from October 3, 2013: http://rogerpielkejr .blogspot.com/2013/10/coverage-of-extreme-events-in-ipcc-ar5.html.

7. BBC History, http://www.bbc.co.uk/history/historic_figures/malthus_thomas .shtml.

8. Christopher Neefus, "Obama Science Czar Called for Carbon Tax to Redistribute Wealth from Global 'North' to 'South,'" CNSNews.com, July 7, 2010, http://cnsnews.com/news/article/obama-science-czar-called-carbon -tax-redistribute-wealth-global-north-south.

9. For a recent example, see Naomi Klein, "Capitalism vs. Climate," *The Nation*, November 28, 2011, http://www.thenation.com/article/164497/capitalism -vs-climate?page=full.

10. See, for instance, Mark Bittman, "The New Nuclear Craze," *New York Times*, August 23, 2013, http://opinionator.blogs.nytimes.com/2013/08/23/the -new-nuclear-craze/?_r=0, in which Bittman says that "Climate change fears should be driving not old and disproven technologies but renewable ones, which are more practical." In the same article, Bittman claims that both nuclear and coal-fired electricity are "doomed."

11. For more on cyberwar, see James Bamford's essay, "The Secret War," *Wired*, June 2013, http://www.wired.com/threatlevel/2013/06/general-keith-alexander -cyberwar/all/.

12. Cliff Saran, "Apollo 11: The computers that put man on the moon," Computerweekly.com, July 2009, http://www.computerweekly.com/feature /Apollo-11-The-computers-that-put-man-on-the-moon.

13. Geoffrey Brumfiel, "Curiosity's Dirty Little Secret," *Slate*, August 20, 2012, http://www.slate.com/articles/health_and_science/science/2012/08/mars_rover _curiosity_its_plutonium_power_comes_courtesy_of_soviet_nukes_.single.html #pagebreak_anchor_2.

14. National Human Genome Research Institute, "DNA Sequencing Costs," undated, http://www.genome.gov/sequencingcosts/.

15. Evgeny Morozov, *To Save Everything Click Here: The Folly of Technological Solutionism* (New York: PublicAffairs, 2013), 357.

16. This calculation is based on daily energy consumption of 250 million barrels per day, with a price of $50 per barrel. At the time of this writing, oil was trading at $100 per barrel. Given that natural gas and coal sell for far less than oil on an energy-equivalent basis, I used $50 as an approximate figure. Thus, on a daily basis, energy-related spending is roughly $12.5 billion. Annually, it comes to $4.56 trillion.

Notes to Chapter 1

1. David McCullough, The Path Between the Seas: The Creation of the Panama Canal 1870–1914 (New York: Simon & Schuster, 1977), 34.

2. Pancanal.com data, http://www.pancanal.com/eng/expansion/rpts/informes -de-avance/expansion-report-201210.pdf, 6.

3. Approximately 5,600 people died during the American construction period. Up to 22,000 died during the French effort. See: pancanal.com, undated, http://www.pancanal.com/eng/general/canal-faqs/index.html.

4. Neil Gershenfeld, "How to Make Almost Anything: The Digital Fabrication Revolution," *Foreign Affairs*, November/December 2012, 48–49.

5. For Culebra Cut volume, see: Panama Canal Museum, "Digging in Hell's Gorge," undated, http://panamacanalmuseum.org/index.php/exhibits/detail/hells_gorge. For total volume of material excavated, see: Frequently asked questions, Canal de Panama, undated, http://www.pancanal.com/eng/general/canal-faqs/index.html.

6. The volume of the stadium is 104 million cubic feet, which is 3.851 million cubic yards. See fact sheet on Cowboys Stadium, http://stadium.dallascowboys.com/assets/pdf/mediaArchitectureFactSheet.pdf. For seating, see: http://stadium.dallascowboys.com/index.html?detectflash=false.

7. David McCullough, *The Path Between the Seas: The Creation of the Panama Canal 1870–1914* (New York: Simon & Schuster, 1977), 544, 547.

8. David McCullough, *The Path Between the Seas: The Creation of the Panama Canal 1870–1914* (New York: Simon & Schuster, 1977), 480.

9. Pancanal.com, undated, http://www.pancanal.com/eng/history/history/index.html.

10. Jonathan Watts, "Nicaragua gives Chinese firm contract to build alternative to Panama Canal," *Guardian*, June 6, 2013, http://www.guardian.co.uk/world/2013/jun/06/nicaragua-china-panama-canal.

11. Pancanal.com, "The French Canal Construction," undated, http://www.pancanal.com/eng/history/history/index.html.

12. David McCullough, *The Path Between the Seas: The Creation of the Panama Canal 1870–1914* (New York: Simon & Schuster, 1977), 167.

13. Ibid.

14. Ibid, 168.

15. The excavation of the Cut stopped in May 1913. David McCullough, *The Path Between the Seas: The Creation of the Panama Canal 1870–1914* (New York: Simon & Schuster, 1977), 604.

16. Pancanal.com, http://www.pancanal.com/eng/history/history/end.html.

17. Wikipedia, http://en.wikipedia.org/wiki/Causes_of_World_War_I.

Notes to Chapter 2

1. For more on Ehsani, see: http://www.ece.tamu.edu/programs/EPI/labs/PEMDL/Main.html.

2. John Medina, *Brain Rules: 12 Principles for Surviving and Thriving at Work, Home, and School* (Seattle, WA: Pear Press, 2008), 39.

3. The sun's luminosity is 3.846^{26} watts. Its mass is 1.98^{30} kg. See: http://nssdc.gsfc.nasa.gov/planetary/factsheet/sunfact.html.

4. For more, see: http://education.jlab.org/qa/plasma_01.html.

5. Steven Johnson, *Where Good Ideas Come From: The Natural History of Innovation* (New York: Riverhead Books, 2010), 46.

6. Francis Bacon, *Novum Organum* (1620), http://wolfweb.unr.edu/homepage/fenimore/engl491/bacon.html.

7. Library of Congress info, http://myloc.gov/exhibitions/bibles/interactives/html/gutenberg/page.html.

8. Abbott Payson Usher, *A History of Mechanical Inventions* (New York: Dover, 1982), 238.

9. The History Guide, "The Printing Press," undated, http://www.historyguide.org/intellect/press.html.

10. P2P Foundation, "Printing Press as an Agent of Change," undated, http://p2pfoundation.net/Printing_Press_as_an_Agent_of_Change.

11. For more, see http://www.gutenberg.org.

12. PBS.org, http://www.pbs.org/transistor/album1/addlbios/deforest.html.

13. For more, see vacuumtubes.net, specifically: http://www.vacuumtubes.net/How_Vacuum_Tubes_Work.html.

14. W. Barksdale Maynard, "Daybreak of the Digital Age," *Princeton Alumni Weekly*, April 4, 2012, http://paw.princeton.edu/issues/2012/04/04/pages/5444/index.xml?page=2&.

15. For the clip from the movie, see: http://movieclips.com/v9zp-the-school-of-rock-movie-the-man/.

16. Gibson.com, http://www2.gibson.com/Products/Electric-Guitars/Les-Paul/Gibson-USA/Les-Paul-Tribute-1952.aspx.

17. Ken Peters, "The Fender Stratocaster: Rock & Roll's Ultimate Design," Nocturnal Design, November 14, 2011, http://www.nocturnaldesign.com/blog/?p=850.

18. Edsullivan.com, http://www.edsullivan.com/artists/the-beatles/.

19. Tim Brookes, *The Guitar: An American Life,* (New York: Grove Press, 2005), 207.

20. Mikhail Safnov, "Confessions of a Soviet moptop," *Guardian*, August 7, 2003, http://www.guardian.co.uk/music/2003/aug/08/thebeatles.

21. Lesliewoodhead.com, http://www.lesliewoodhead.com/2011/05/how-the-beatles-rocked-the-kremlin-the-book/.

22. Documentary is available on WNET: http://www.thirteen.org/beatles/video/video-watch-how-the-beatles-rocked-the-kremlin/.

23. Amazon data, http://www.amazon.com/How-Beatles-Rocked-Kremlin-Revolution/dp/1608196143/ref=sr_1_1?ie=UTF8&qid=1363229791&sr=8-1&keywords=How+the+Beatles+Rocked+the+Kremlin.

24. Toni O'Laughlin, "Truth after 42 years: Beatles banned for fear of influence on youth," *Guardian*, September 21, 2008, http://www.guardian.co.uk/world/2008/sep/22/israelandthepalestinians.thebeatles.

25. James Shingler, "Rocking the Wall: East German Rock and Pop in the 1970s and 1980s," undated, http://thevieweast.wordpress.com/2011/07/15/rocking-the-wall-east-german-rock-and-pop-in-the-1970s-and-1980s/.

26. Virginmedia.com, http://www.virginmedia.com/music/pictures/toptens/banned-popstars.php?ssid=6.

27. Eric R. Danton, "Pussy Riot Member: 'What Happened to Us is Unacceptable,'" Rollingstone.com, March 25, 2013, http://www.rollingstone.com/music/videos/pussy-riot-member-what-happened-to-us-is-unacceptable-20130325.

28. PBS.org, http://www.pbs.org/transistor/album1/addlbios/deforest.html.

29. Loz Blain, "Inventions that changed the world: Mikhail Kalashnikov's AK-47," July 22, 2009, http://www.gizmag.com/kalashnikov-ak-47/12306/.

30. C. J. Chivers, *The Gun*, excerpted in Esquire, October 27, 2010, http://www.esquire.com/features/ak-47-history-1110.

31. For more, see: http://www.gunclassics.com/list.html.

32. Amnesty International, "TheAK-47: The World's Favorite Killing Machine," June 26, 2006, http://www.amnesty.org.uk/uploads/documents/doc_17010.pdf, 4.

33. Ibid.

34. See, for instance, "Ak74 Torture test," May 31, 2011, http://www.youtube.com/watch?v=VHwDfx5nCrA.

35. BBC, "AK-47: Iconic Weapon," December 5, 2005, http://news.bbc.co.uk/2/hi/4380348.stm.

36. Flag data on Mozambique available here: http://www.mapsofworld.com/images/world-countries-flags/mozambique-flag.gif.
The Hezbollah flag can be seen here: http://en.wikipedia.org/wiki/File:Hezbollah_Flag.jpg.

37. Nicholas Schmidle, "Top Gun," *Slate*, November 1, 2010, http://www.slate.com/articles/arts/books/2010/11/top_gun.single.html#pagebreak_anchor_2.

38. For more on this, see globalissues.org, http://www.globalissues.org/print/article/78#UNConferenceontheIllicitTradeinSmallArmsJuly2001.

39. BBC, "AK-47: Iconic Weapon," December 5, 2005, http://news.bbc.co.uk/2/hi/4380348.stm.

40. Vaclav Smil, *Creating the Twentieth Century: Technical Innovations of 1867–1914 and Their Lasting Impact* (Oxford: Oxford University Press, 2005), 186.

41. Nobelprize.org, http://www.nobelprize.org/nobel_prizes/chemistry/laureates/1918/.

42. Bosch's award was for his contributions to the invention and development of chemical high pressure methods, which he shared with Friedrich Bergius. See: http://www.nobelprize.org/nobel_prizes/chemistry/laureates/1931/index.html.

43. Earth Policy Institute, http://www.earth-policy.org/data_center/C24.

44. Smil, *Creating the Twentieth Century*, 195.

45. Ibid., 196.

46. Vaclav Smil, *Prime Movers of Globalization: The History and Impact of Diesel Engines and Gas Turbines,* (Cambridge, MA: MIT Press, 2010), 18.

47. Ibid, 19.

48. Diesel Technology Forum, press release, March 21, 2012, http://www.diesel forum.org/news/-president-obama-a-successful-all-of-the-above-energy-policy -must-not-seek-to-pick-winners-and-losers-allen-schaeffer-diesel-technology -forum.

49. Diesel Technology Forum, "Diesel Powers the U.S. Economy," undated, http:// www.dieselforum.org/files/dmfile/DTF_economrpt_full.pdf, 4.

50. Diesel's energy density is about 147,000 Btu per gallon. Gasoline contains about 125,000 Btu per gallon.

51. Smil, *Prime Movers of Globalization,* 49.

52. Ibid., 47.

53. Ibid., 62.

54. For more on this company, see my last book, *Power Hungry*: The Myths of "Green" Energy and the Real Fuels of the Future (New York: PublicAffairs, 2010), 43.

55. Vaclav Smil, "The Two Prime Movers of Globalization," *Journal of Global History* (2007): 377–378.

56. Ibid, 391.

57. International Air Travel Association, "2012 Annual Review," June 2012, http://www.iata.org/about/Documents/annual-review-2012.pdf, 10.

58. Edwin McDowell, "Business Travel; Air Fares to Europe are Cheap and May Stay Low for a While," *New York Times*, February 17, 1999, http://www.nytimes .com/1999/02/17/business/business-travel-air-fares-to-europe-are-cheap-and -may-stay-low-for-a-while.html?pagewanted=all&src=pm.

59. Using an inflation calculator for 2011 (http://www.westegg.com/inflation /infl.cgi), the sum was $7,700.

60. United-states-lines.org, "Remembering the Days of Civilized Air Travel," undated, http://united-states-lines.org/Pam%20Am%20Stratocruiser.html.

61. Smil, *Prime Movers of Globalization,* 79.

62. Ibid., 82.

63. Airlines for America data, http://www.airlines.org/Pages/Annual-Round-Trip -Fares-and-Fees-Domestic.aspx.

64. For more on the B-29, see Boeing's website: http://www.boeing.com/history /boeing/b29.html. For the density of the Wright engine, see: http://en.wikipedia .org/wiki/Wright_Flyer#Specifications_.28Wright_Flyer.29.

65. Image available here: http://www.airlinereporter.com/wp-content/uploads /2012/07/GEnx-1B-with-person-d42657B.jpg.

66. GE Aviation data, http://www.geaviation.com/engines/commercial/genx/.

67. In 2011, 2.8 billion people traveled on commercial airlines. That works out to 7.6 million people per day. By 2016, the IATA expects 3.6 billion passengers. See: International Air Transport Association, "Airlines to Welcome 3.6 Billion Passengers in 2016," December 6, 2012, http://www.iata.org/pressroom/pr/pages/2012–12–06–01.aspx.

68. Debora MacKenzie, "Haiti Caught Cholera from UN Peacekeepers," NewScientist.com, May 6, 2011, http://www.newscientist.com/blogs/shortsharpscience/2011/05/haiti-caught-cholera.html.

69. Mark Doyle, "Haiti Cholera Epidemic 'Most Likely' Started at UN Camp–Top Scientist," BBC.co.uk, October 22, 2012, http://www.bbc.co.uk/news/world-latin-america-20024400.

70. For more on the Hobby-Eberly, see Robert Bryce, "Frugal Texans Build Cutting-Edge Telescope," *Christian Science Monitor*, November 22, 1996, http://www.csmonitor.com/1996/1122/112296.us.us.2.html. For Manchester device, see Dailymail.co.uk, "World's most powerful optical microscope 'could view live viruses,'" March 1, 2011, http://www.dailymail.co.uk/sciencetech/article-1361863/Most-powerful-optical-microscope-world-soon-view-live-viruses.html.

71. James E. McClellan III and Harold Dorn, *Science and Technology in World History: An Introduction* (Baltimore: Johns Hopkins University Press, 2006), 224.

72. Wikipedia.org, http://en.wikipedia.org/wiki/Sidereus_Nuncius.

73. Arthur Koestler, *The Sleepwalkers: A History of Man's Changing Vision of the Universe* (London: Penguin Books, 1959), 356.

74. Alan Cowell, "After 350 Years, Vatican Says Galileo Was Right: It Moves," *New York Times*, October 31, 1992, http://www.nytimes.com/1992/10/31/world/after-350-years-vatican-says-galileo-was-right-it-moves.html.

75. http://www.scitechantiques.com, "Galileo's Original Telescopes: New measurements of their dimensions with special optical rulers, undated, http://galileo.rice.edu/sci/instruments/telescope.html.

76. Rice.edu, The Galileo Project, "The Telescope," undated, http://galileo.rice.edu/sci/instruments/telescope.html.

77. Vision Engineering, "Ancient History," undated, http://www.visioneng.com/history-of-the-microscope.php.

78. For more, see, for instance, opticsplanet.com.

79. Wikipedia, http://en.wikipedia.org/wiki/Thomas Edison.

80. Rutgers.edu, "Edison's Patents," undated, http://edison.rutgers.edu/patents.html.

81. Kevin Maney, "Search for the Most Prolific Inventors Is a Patent Struggle," *USA Today*, December 6, 2012, http://usatoday30.usatoday.com/money/industries/technology/maney/2005–12–06–top-patent-hoders_x.htm?goback=%252Egde_3718887_member_39658477.

82. Baseball-almanac.com, http://www.baseball-almanac.com/feats/feats3.shtml; Nba.com data, http://www.nba.com/sixers/news/wilt_boxscore.html.

83. David E. Nye, *Electrifying America: Social Meanings of a New Technology* (Cambridge: MIT Press, 1992), 2.

84. Ibid., 225.

85. Ibid., 159.

86. Smil, *Creating the Twentieth Century*, 89.

87. Rakteem Katakey & Winnie Zhu, "Coal 4-Year Low Lures Utilities Ignoring Climate: Energy Markets," *Bloomberg*, October 11, 2013, http://www.bloomberg .com/news/2013–10–11/coal-4-year-low-lures-utilities-ignoring-climate-energy -markets.html.

88. The paper is still operating today. Its tag line is "Shining light on Wabash County since 1859." For more, see: http://www.chronicle-tribune.com/wabash plaindealer/.

89. Nye, *Electrifying America,* 3.

90. US Patent Office, patent number 930,759. See: http://www.google.com/patents /US930759.

91. Ford data, http://media.ford.com/article_display.cfm?article_id=858.

92. American Society of Mechanical Engineers, "Hughes Two-Cone Drill Bit," August 10, 2009, http://files.asme.org/MEMagazine/Web/20779.pdf, 1.

93. Ibid., 5.

94. EIA data, http://www.eia.gov/dnav/pet/hist/LeafHandler.ashx?n=PET&s =MCRFPUS2&f=A.

95. Ibid, 2.

96. Fishtail bit from the Houston Museum of Natural Science, http://commons .wikimedia.org/wiki/File:Fishtail_Bit_(drag_bit)_-_Houston_Museum_of _Natural_Science_-_DSC01337.JPG.

97. See: http://www.reformation.org/saint-martin-luther.html.

98. Anthony Shadid, "Syria's Sons of No One," *New York Times Magazine*, August 31, 2011, http://www.nytimes.com/2011/09/04/magazine/syrias-sons -of-no-one.html?pagewanted=all.

99. Bruce Schneier, Crypto-gram email, April 15, 2013.

100. Peter Maass and Megha Rajagopalan, "That's No Phone. That's My Tracker," *New York Times*, July 13, 2012, http://www.nytimes.com/2012/07/15/sunday -review/thats-not-my-phone-its-my-tracker.html?_r=2&.

101. *Economist*, "Look Who's Listening," June 15, 2013, http://www.economist .com/news/briefing/21579473-americas-national-security-agency-collects -more-information-most-people-thought-will.

102. Alexis Madrigal, "I'm Being Followed: How Google—and 104 Other Companies—are Tracking Me on the Web," the *Atlantic,* February 29, 2012, http://

www.theatlantic.com/technology/archive/2012/02/im-being-followed-how
-google-151-and-104-other-companies-151-are-tracking-me-on-the-web
/253758/.

103. Steven Pinker, *The Better Angels of Our Nature: Why Violence Has Declined*
(New York: Viking, 2011), 478.

Notes to Chapter 3

1. Abdel R. Omran, "The Epidemiologic Transition: A Theory of the Epide-
miology of Population Change," *Milbank Memorial Fund Quarterly* 49, no. 4, pt.
1, (1971): 509–538. http://pingpong.ki.se/public/pp/public_courses/course07443
/published/0/resourceId/0/content/20.11%20The%20Epidemiologic%20Transition
.pdf.

2. US Census Bureau, "Health, United States, 2010," Table 22, http://www.cdc
.gov/nchs/data/hus/hus10.pdf#022.

3. UNICEF data, http://www.unicef.org/sowc2013/files/Table_6_Stat_Tables
_SWCR2013_ENGLISH.pdf.

4. World Health Organization, World Health Statistics, Part III, Global health
indicators, Table 1, 61, http://www.who.int/healthinfo/EN_WHS2012_Part3
.pdf.

5. WHO, World Health Statistics 2012, 13, http://www.who.int/gho/publications
/world_health_statistics/EN_WHS2012_Full.pdf.

6. Matthew Ridley, "Apocalypse Not," *Wired*, August 17, 2012, http://www.wired
.com/wiredscience/2012/08/ff_apocalypsenot/all/.

7. Gregg Easterbrook, *The Progress Paradox: How Life Gets Better While People
Feel Worse* (New York: Random House, 2003), 69.

8. UNESCO, UNESCO Institute for Statistics, "Adult and Youth Literacy,"
September 2011, http://www.uis.unesco.org/FactSheets/Documents/FS16
–2011-Literacy-EN.pdf.

9. http://www.uis.unesco.org/literacy/Pages/adult-youth-literacy-data-viz
.aspx.

10. Data from UNESCO shows a close correlation between wealth and literacy.
The countries that have per-capita GDP of about $5,000 or more per year have
literacy rates approaching 100 percent. Thus as poverty declines, we can assume
that literacy increases.

11. John Mackey and Raj Sisodia, *Conscious Capitalism: Liberating the Heroic Spirit
of Business* (Boston: Harvard Business Review Press, 2013), 12.

12. World Bank data, http://web.worldbank.org/WBSITE/EXTERNAL
/TOPICS/EXTPOVERTY/EXTPA/0,,contentMDK:20040961~menu
PK:435040~pagePK:148956~piPK:216618~theSitePK:430367~isCURL:Y,00
.html.

13. Maxim Pinkovskiy and Xavier Sala-i-Martin, "Parametric Estimations of the World Distribution of Income," National Bureau of Economic Research, October 2009, http://www.nber.org/papers/w15433.pdf, 1.

14. http://www.freedomhouse.org/sites/default/files/inline_images/FIW%20 2012%20Booklet—Final.pdf, p. 29.

15. Steven Pinker, *The Better Angels of Our Nature: Why Violence Has Declined* (New York: Viking, 2011), xxi.

14. Pinkovskiy and Martin, "Parametric Estimations of the World Distribution of Income," Figure 24, 53.

17. Richard Heinberg, *Peak Everything: Waking up to the Century of Declines* (Gabriola Island, BC, Canada: New Society, 2007).

18. John Boyce, "Peak Oil: Mountain or Molehill? An Empirical Assessment," November 2010, 30. Boyce published an expanded version of this paper in July 2012. See: https://webdisk.ucalgary.ca/~boyce/public_html/Peak%20Oil%20Paper %20Final%20LaTeX%20version.pdf.

19. *The Economist,* "Crowded Out," September 24, 2011, http://www.economist .com/node/21528986.

20. National Renewable Energy Laboratory, "2010 Solar Technologies Market Report," November 2011, http://www.nrel.gov/docs/fy12osti/51847.pdf, 60.

21. EIA data, http://www.eia.gov/renewable/annual/solar_photo/pdf/pv_report .pdf, 3.

22. Eric Wesoff, "First Solar Surprises with Big 2013 Guidance, 40 Cents per Watt Cost by 2017," Greentechmedia.com, April 9, 2013, http://www.green techmedia.com/articles/read/First-Solar-Surprises-With-Big-2013-Guidance-40 -Cents-Per-Watt-Cost-by-201.

23. BP Statistical Review of World Energy 2013. Throughout this book, I refer to BP data that I downloaded in an Excel spreadsheet, which I then used to calculate various percentages and growth rates. The original data can be downloaded here: http://www.bp.com/en/global/corporate/about-bp/statistical -review-of-world-energy-2013.html.

24. National Renewable Energy Laboratory, "2010 Solar Technologies Market Report," November 2011, http://www.nrel.gov/docs/fy12osti/51847.pdf, 60.

25. Bureau of Transportation Statistics data.

26. Countdowntokittyhawk.com, undated, http://www.countdowntokitty hawk.com/news/031017_fact_sheet.html.

27. Andy Pasztor, "Virgin Galactic Spacecraft Tests Rocket in Flight," *Wall Street Journal,* April 29, 2013, http://online.wsj.com/article/SB1000142412788732 3798104578452901565132188.html.

28. Bureau of Transportation Statistics data.

29. US EPA data, http://cfpub.epa.gov/eroe/index.cfm?fuseaction=detail.view Ind&lv=list.listByAlpha&r=188208&subtop=341.

30. US EPA data, http://cfpub.epa.gov/eroe/index.cfm?fuseaction=detail.view Ind&lv=list.listbyalpha&r=219697&subtop=341.

31. BP Statistical Review of World Energy 2013. In 1990, US energy consumption was 1.97 billion tons of oil equivalent. In 2005, it was 2.35 billion tons of oil equivalent.

32. Earth Policy Institute, "World Grain Yields, Annual Percent Increase by Decade, 1950–2010," http://www.earth-policy.org/datacenter/xls/book_wote_ch12 _4a.xls.

33. Sarah Brown, "London 2012: The women's Olympics?" CNN.com, August 10, 2012, http://edition.cnn.com/2012/08/10/sport/london-olympics-women.

34. Wikipedia.org, http://en.wikipedia.org/wiki/Women's_suffrage.

35. Cooperative Institutional Research Program, "The American Freshman: National Norms Fall 2012," undated (approximately January 2013), http://heri.ucla .edu/monographs/TheAmericanFreshman2012.pdf, 51.

Notes to Chapter 4

1. Michael Shellenberger and Ted Nordhaus, "Evolve: A Case for Modernization as the Road to Salvation," *Orion*, September/October 2011, http://www.orion magazine.org/index.php/articles/article/6402/.

2. David Deming, "What the Oil Business Could Learn from the NRA," *Wall Street Journal*, March 1, 2013, A11

3. Pascal Bruckner, "The Ideology of Catastrophe," *Wall Street Journal*, April 10, 2012.

4. See these two posts on Gore's blog in September 2013: http://blog.algore.com /2013/09/.

5. See: http://blog.algore.com/2012/09/planetary_emergency.html.

6. Greenpeace.org, http://www.greenpeace.org/international/en/about/reports /#ao.

7. Rex Weyler, "Deep Green: Why De-Growth? An Interview," Greenpeace .org, June 27, 2011, http://www.greenpeace.org/international/en/news/Blogs /makingwaves/deep-green-why-de-growth-an-interview/blog/35467/.

8. Naomi Klein, "Capitalism vs. the Climate," *The Nation*, November 11, 2011, http://www.thenation.com/article/164497/capitalism-vs-climate?page=full.

9. Worldwatch Institute, "The Path to Degrowth in Overdeveloped Countries," April 2012, http://blogs.worldwatch.org/sustainableprosperity/wp-content /uploads/2012/04/Chapter-2-Summary.pdf.

10. Ibid.

11. John M. Broder, "Obama's Remarks Offer Hope to Opponents of Oil Pipeline," *New York Times*, July 5, 2013, http://www.nytimes.com/2013/07/06/us /obamas-remarks-offer-hope-to-opponents-of-oil-pipeline.html?src=recg.

12. McKibben quoted by Matthew C. Nisbet, "The Opponent," *Options Politiques*, April–May 2013, http://www.irpp.org/assets/po/arctic-visions/nisbet.pdf. Quote is from McKibben's 2007 book, *Deep Economy*.

13. Bill McKibben, "Global Warming's Terrifying New Math," *Rolling Stone,* July 19, 2012, http://www.rollingstone.com/politics/news/global-warmings-terrifying-new-math-20120719?print=true.

14. McKibben quoted by Matthew C. Nisbet, "The Opponent," *Options Politiques,* April–May 2013, http://www.irpp.org/assets/po/arctic-visions/nisbet.pdf.

15. Atlantic, "How and When Will the World End," June 19, 2013, http://www.the atlantic.com/magazine/archive/2013/07/how-and-when-will-the-world-end-/309400/.

16. McKibben, *Eaarth: Making a Life on a Tough New Planet* (Toronto: Vintage, 2010), 184, http://books.google.com/books?id=wwbwUDpoPrkC&pg=PA184&lpg=PA184&dq="to+cut+our+fossil+fuel+use+by+a+factor+of+twenty+over+the+next+few+decades."&source=bl&ots=uO4fXn7lcI&sig=TWEboCktfvRtVdJzcQa6zgBdukE&hl=en&sa=X&ei=-9ybULjWMsPmygHG3oGQCQ&ved=oCEsQ6AEwBg#v=onepage&q&f=false.

17. BP Statistical Review of World Energy 2013.

18. BP Statistical Review of World Energy 2011.

19. Energy Information Administration data, http://www.eia.gov/cfapps/ipdb project/iedindex3.cfm?tid=5&pid=62&aid=2&cid=ww,&syid=2009&eyid=2010 &unit=TBPD.

20. BP Statistical Review of World Energy 2013. Total hydrocarbon use in 2011 averaged 215 million barrels of oil equivalent per day. That equals 34.1 billion liters. Divided by 7 billion people equals 4.88 liters/person.

21. This assumes the global average of 87 percent of all energy being produced from hydrocarbons.

22. BP Statistical Review of World Energy 2013.

23. One liter is equal to 0.26 gallons. The 2013 Prius gets 50 miles per gallon. A standard 2013 Suburban 1500 with two-wheel drive gets 17 mpg. See: http://www.fueleconomy.gov/feg/Find.do?action=sbs&id=33324&id=32470.

24. Bill McKibben and Marlene Spoerri, "Ethics Matter: A Conversation with Bill McKibben," *Policy Innovations,* October 16, 2012, http://www.policyinnovations.org/ideas/audio/data/000644.

25. BP Statistical Review of World Energy 2013.

26. John Noble Wilford, "Don't Blame Columbus for All the Indians' Ills," *New York Times,* October 29, 2002, http://www.nytimes.com/2002/10/29/science/don-t-blame-columbus-for-all-the-indians-ills.html?pagewanted=all&src=pm.

27. Pascal Bruckner, *The Fanaticism of the Apocalypse* (Malden, MA: Polity Press, 2013), 145.

28. Encyclopedia Britannica, "noble savage," undated, http://www.britannica.com/EBchecked/topic/416988/noble-savage.

29. Rousseau wrote the essay in 1754. It was published in 1755. See: http://en.wikipedia.org/wiki/Discourse_on_Inequality.

30. Jean-Jacques Rousseau, "A Discourse upon the Origin and the Foundation of the Inequality Among Mankind," 1755, Full text available: http://www.gutenberg.org/cache/epub/11136/pg11136.txt.

31. Jennifer Schuessler, "Thoreau's Pencil," *New York Times*, May 1, 2009, http://artsbeat.blogs.nytimes.com/2009/05/01/thoreaus-pencil/.

32. From *Walden*, cited by Randy Alfred, "Aug. 9, 1854: Thoreau Warns, 'The Railroad Rides on Us,'" Wired.com, August 9, 2010, http://www.wired.com/thisdayintech/2010/08/0809thoreau-walden-published/.

33. Henry David Thoreau, *Walden*, conclusion, full text here: http://xroads.virginia.edu/~hyper/walden/hdt18.html.

34. John Updike, "A sage for all seasons," *The Guardian*, June 25, 2004, http://www.theguardian.com/books/2004/jun/26/classics.

35. Dorothy McLaughlin, "Silent Spring Revisited," undated, http://www.guardian.co.uk/books/2004/jun/26/classics; http://www.pbs.org/wgbh/pages/frontline/shows/nature/disrupt/sspring.html. For quote, see: Rachel Carson, *Silent Spring*. Excerpt available here: http://core.ecu.edu/soci/juskaa/SOCI3222/carson.html.

36. Donella H. Meadows, Dennis L. Meadows, Jorgen Randers, and William W. Behrens III, *The Limits to Growth: A Report for The Club of Rome's Project on the Predicament of Mankind* (New York: Universe Books, 1972), http://web.ics.purdue.edu/~wggray/Teaching/His300/Illustrations/Limits-to-Growth.pdf.

37. Edward Abbey, *Beyond the Wall: Essays from the Outside* (New York: Henry Holt, 1971), 40.

38. Wikipedia.org, http://en.wikipedia.org/wiki/The_Population_Bomb.

39. Cited by Sarah Orleans Reed in "The Publication of Paul Ehrlich's *The Population Bomb* by the Sierra Club, 1968: Wilderness-Thinking, Neo-Malthusianism, and Anti-Humanism," Wesleyan University, April 2008, 5.

40. Alan Gregg, "A Medical Aspect of the Population Problem," *Science* 121 (1955): 681–682, p. 682, cited in Charles Rubin, "Human Dignity and the Future of Man," President's Council on Bioethics, March 2008, http://bioethics.georgetown.edu/pcbe/reports/human_dignity/chapter7.html#endnote25.

41. Huffingtonpost.com, "Sir David Attenborough, Naturalist and Filmmaker, Calls Humanity 'A Plague on the Earth,'" January 23, 2013, http://www.huffingtonpost.com/2013/01/23/sir-david-attenborough-na_n_2534078.html?utm_hp_ref=tw#slide=1409005.

42. Bruckner, "The Ideology of Catastrophe."

43. George Monbiot, "Is There Any Point in Fighting to Stave Off Industrial Apocalypse?" *Guardian*, August 17, 2009, http://www.guardian.co.uk/commentisfree/cif-green/2009/aug/17/environment-climate-change.

44. Will Steffen, Johan Rockström, and Robert Costanza, "How Defining Planetary Boundaries Can Transform Our Approach to Growth," *Solutions Journal,* May 2011, http://www.thesolutionsjournal.com/node/935?page=35%2C02C0% 2C0%2C0%2C0.

45. Http://www.sierraclub.org/policy/conservation/nuc-power.aspx.

Notes to Chapter 5

1. Richard Dorment, "Eadweard Muybridge at Tate Britain, Review," *Telegraph,* September 8, 2010, http://www.telegraph.co.uk/culture/art/art-reviews/7985248 /Eadweard-Muybridge-at-Tate-Britain-review.html.

2. University of Ottawa press release, http://www.uottawa.ca/articles/freeze -frame-physics.

3. For more, see: http://www.cfa.harvard.edu/itamp/attosecond/attosecond.html.

4. PBS.org, "Watson and Crick Describe Structure of DNA: 1953," undated, http://www.pbs.org/wgbh/aso/databank/entries/do53dn.html.

5. Wikipedia.org, "electromagnetic spectrum," undated, http://en.wikipedia.org /wiki/Electromagnetic_spectrum.

6. Nobelprize.org, http://www.nobelprize.org/nobel_prizes/medicine/laureates /1962/press.html.

Notes to Chapter 6

1. Guinnessworldrecords.com data, http://www.guinnessworldrecords.com /world-records/speed/fastest-100m-with-a-can-balanced-on-head-(dog).

2. Guinnessworldrecords.com data, http://www.guinnessworldrecords.com /world-records/speed/fastest-100-metre-hurdles-wearing-swim-fins-(individual -male).

3. Guinnessworldrecords.com data, http://www.guinnessworldrecords.com /world-records/speed/fastest-time-to-hula-hoop-10-km-(male).

4. Http://library.thinkquest.org/C004203/science/science02.html.

5. Endless Sphere Technologies, diameter to speed/rpm chart. http://endless -sphere.com/forums/viewtopic.php?f=28&t=16114.

6. Http://www.vanderbiltcupraces.com/drivers/bio/robert_burman.

7. Holman W. Jenkins Jr., "Jenkins: Saying 'Yes' to Broadband," *Wall Street Journal,* October 9, 2013, http://online.wsj.com/news/articles/SB1000142405270 2304626104579123202528096562.

8. Vaclav Smil, *Prime Movers of Globalization: The History and Impact of Diesel Engines and Gas Turbines* (Cambridge, MA: MIT Press, 2010), 12.

9. Max Kingsley-Jones, "6,000 and Counting for Boeing's Popular Little Ttwin jet," April 22, 2009, http://www.flightglobal.com/news/articles/pictures-6000 -and-counting-for-boeings-popular-little-twinjet-325472/.Boeing speed data:

http://www.boeing.com/commercial/737family/pf/pf_800tech.html Conversion from Mach to km/h and mph here: http://www.globalaircraft.org/converter .html.

10. Enchantedlearning.com, "Christopher Columbus: Explorer," undated, http:// www.enchantedlearning.com/explorers/page/c/columbus.shtml.

11. BBC.co data, http://www.bbc.co.uk/england/sevenwonders/west/ss_gb _mm/index.shtml.

12. Sarah Hoye, "Sending Out an SOS for 'America's Flagship,'" CNN.com, April 7, 2013, http://www.cnn.com/2013/04/07/travel/ocean-liner-united-states; John Steele Gordon, "The Anti-Titanic," *Wall Street Journal,* August 17, 2012, http://online.wsj.com/article/SB10000872396390443991704577579623942024142 .html?mod=googlenews_wsj.

13. Great short film on the Fosbury flop here: http://www.youtube.com/watch ?v=Z_sIwv6SAxc.

14. Tufts.edu, "Ancient Olympic Events," undated, http://www.perseus.tufts .edu/Olympics/running.html.

15. Marathonguide.com, "Marathoners Become Ever Faster," undated, http:// www.marathonguide.com/history/records/index.cfm.

16. Kevin Quealy and Graham Roberts, "Usain Bolt vs. 116 Years of Olympic Sprinters," *New York Times*, August 5, 2012, http://www.nytimes.com/interactive /2012/08/05/sports/olympics/the-100-meter-dash-one-race-every-medalist -ever.html?smid=fb-shar.

17. Theweek.com, "Usain Bolt's Prodigious Speed: By the Numbers," August 9, 2012, http://theweek.com/article/index/231813/usain-bolts-prodigious-speed -by-the-numbers.

18. Matthew Futterman, Jonathan Clegg, and Geoffrey A. Fowler, "An Olympics Built for Records," *Wall Street Journal*, August 9, 2012, http://online.wsj.com /article/SB10000872396390443991704577579320489432242.html.

19. Http://www.roadsters.com/750/.

20. The leader of the Bloodhound project is Richard Noble, see: http://www .bloodhoundssc.com. He also led the Thrust SSC, http://www.thrustssc.com /thrustssc/contents_frames.html.

Notes to Chapter 7

1. Roger Pielke Jr., "Looking into the Soul of Sport," The Least Thing blog, October 15, 2012, http://leastthing.blogspot.com/2012/10/looking-into-soul -of-sport.html?spref=tw.

2. Tyler Hamilton and Daniel Coyle, *The Secret Race: Inside the Hidden World of the Tour de France: Doping, Cover-ups, and Winning at All Costs* (New York: Bantam Books, 2012), 185.

3. Ibid., 104.

4. *Daily Telegraph*, "Lance Armstrong's Doctor Michele Ferarri Denies Seeing Disgraced Cyclist Use Performance-Enhancing Drugs," December 14, 2012, http://www.telegraph.co.uk/sport/othersports/cycling/lancearmstrong/9746608 /Lance-Armstrongs-doctor-Michele-Ferrari-denies-seeing-disgraced-cyclist -use-performance-enhancing-drugs.html.

5. Hamilton and Coyle, *The Secret Race*, 104.

6. James Dao, "Watchdogs Seek Doping Clues from a Distance," *New York Times*, July 17, 2013, http://www.nytimes.com/2013/07/18/sports/cycling/during -tour-de-france-watchdogs-seek-doping-clues-from-a-distance.html?ref=sports &pagewanted=all.

7. This data is from Austin-based amateur racer Frank Kurzawa, who continually monitors his power output while riding.

8. The Editors of Bicycling Magazine, *The Noblest Invention: An Illustrated History of the Bicycle* (United States: Rodale, 2003), 272.

9. James L. Witherell, *Bicycle History: A Chronological Cycling History of People, Races, and Technology* (Cherokee Village, AR: McGann, 2010), 46.

10. Ibid., 50.

11. FelixWong.com, "Tour de France Bicycles & Historical Bike Weights," November 24, 2010, http://felixwong.com/2010/11/tour-de-france-bicycles -historical-bike-weights/.

12. Witherell, op. cit., 114.

13. FelixWong.com, "Tour de France Bicycles & Historical Bike Weights," November 24, 2010, http://felixwong.com/2010/11/tour-de-france-bicycles -historical-bike-weights/.

14. Ben Hewitt, "Tour de Lance," *Wired*, July 2004, http://www.wired.com/wired /archive/12.07/armstrong.html.

15. Ibid.

16. For more on this see mapawatt.com. "How May Watts Can you Produce?" July 19, 2009, mapawatt.com/2009/07/19/bicycle-power-watts/.

17. James Huang, "Tour de France Winning Bikes," Bikeradar.com, June 29, 2012, http://www.bikeradar.com/road/news/article/tour-de-france-winning -bikes-34375/.

18. Competitive Cyclist data, http://www.competitivecyclist.com/product -components/2012-Easton-EC90-SL-Carbon-Clincher-Wheelset-8815.41.1 .html.

19. James L. Witherell, *Bicycle History,* 93.

20. Http://cyclinginfo.co.uk/blog/450/procycling/cycling-doping-scandals/.

21. James L. Witherell, *Bicycle History,* 150.

22. BBC.co.uk, "Jan Ullrich: Former Tour de France Winner Admits Blood Doping," June 22, 2013, http://www.bbc.co.uk/sport/0/cycling/23013133.

23. *Guardian*, "The Festina Affair," July 9, 2008, http://www.guardian.co.uk /sport/gallery/2008/jul/09/tourdefrance.cycling.

24. Oprah Winfrey, "Oprah Talks to Lance Armstrong," *O, The Oprah Magazine*, May 2004.

25. Bonnie D. Ford, "Landis Admits Doping, Accuses Lance," ESPN.com, May 21, 2010, http://sports.espn.go.com/oly/cycling/news/story?id=5203604.

26. Jonathan Vaughters, "How to Get Doping Out of Sports," *New York Times*, August 12, 2012, 9.

27. CNN, "Lance Armstrong's Epic Downfall," October 22, 2012, http://edition .cnn.com/2012/10/22/sport/lance-armstrong-controversy/index.html.

28. David Bauder, "Lance to Oprah: Yes, Yes, Yes, Yes, and Yes," Associated Press, January 18, 2013, http://nbcsports.msnbc.com/id/50508056/ns/sports -cycling/.

Notes to Chapter 8

1. Donald Hill, A History of Engineering in Classical and Medieval Times (London: Routledge, 1984), 145–146.

2. Kerry Harrison, "Irrigation Pumping Plants and Energy Use," University of Georgia College of Agricultural and Environmental Sciences, undated, http:// www.caes.uga.edu/publications/pubDetail.cfm?pk_id=6025.

3. Humandynamo.com, http://www.humandynamo.com/technical_info.html.

4. Vaclav Smil, *Energies: An Illustrated Guide to the Biosphere and Civilization* (Cambridge, MA: Massachusetts Institute of Technology, 1999) 112.

5. Ibid., 117.

6. Ibid., 120–121.

7. Ibid., 123.

8. Mario Pozner, "Illustrated History of Wind Power Development," undated, http://telosnet.com/wind/early.html.

9. Thomas Content, "State's Tallest Wind Turbines to Begin Operating near Green Bay," *Journal Sentinel*, November 10, 2010, http://www.jsonline.com/business /107011438.html.

10. Smil, *Energies*, 125.

11. Charles R. Morris, *The Dawn of Innovation: The First American Industrial Revolution* (New York: PublicAffairs, 2012), 100–103.

12. Ibid., 102.

13. Ibid., 43.

14. Performance Trends Inc., "Cylinder Pressure," undated, http://performance trends.com/Definitions/Cylinder-Pressure.html.

15. Http://simple.wikipedia.org/wiki/File:James_Watt_by_Henry_Howard.jpg.

16. William Rosen, *The Most Powerful Idea in the World: A Story of Steam, Industry, and Invention* (New York: Random House, 2010), 103.

17. Ibid., 160.

18. Morris, *The Dawn of Innovation*, 43–44.

19. Carl Lira, "Biography of James Watt," undated, http://www.egr.msu.edu/~lira/supp/steam/wattbio.html.

20. PBS.org, "Who Made America?" Profile of Fulton, undated, http://www.pbs.org/wgbh/theymadeamerica/whomade/fulton_hi.html.

21. Invent.org, Fulton profile, undated, http://www.invent.org/hall_of_fame/268.html.

22. Michelle and James Nevius, "Robert Fulton and the Age of Steam," Inside theApple.net, August 17, 2010, http://blog.insidetheapple.net/2010/08/robert-fulton-and-age-of-steam.html.

23. New York State Education Department, "Steamboats on the Hudson: An American Saga," undated, http://www.nysl.nysed.gov/mssc/steamboats/amercit4.html.

24. Morris, op. cit., 175–176.

25. Jeff Goodell, *Big Coal: The Dirty Secret Behind America's Energy Future* (Boston: Houghton Mifflin, 2006), 75

26. Rosen, *The Most Powerful Idea in the World*, xxii.

27. Goodell, op. cit., 75.

28. Alex Epstein, "Vindicating Capitalism: The Real History of the Standard Oil Company," *The Objective Standard*, Summer 2008, http://www.theobjectivestandard.com/issues/2008-summer/standard-oil-company.asp#_edn22.

29. Ron Chernow, *Titan: The Life of John D. Rockefeller, Sr.* (New York: Vintage Books, 1998), 113.

30. Nicole Mordant, "Analysis: Crude-by-Rail Carves Out Long-term North American Niche," Reuters, November 4, 2012, http://www.reuters.com/article/2012/11/04/us-railways-oil-northamerica-idUSBRE8A30AX20121104.

31. Adnan Vatansever, "Russia's Oil Exports: Economic Rationale Versus Strategic Gains," Carnegie Endowment for International Peace, December 2010, http://carnegieendowment.org/2010/12/15/russia-s-oil-exports-economic-rationale-versus-strategic-gains/oza.

32. Morris, *The Dawn of Innovation*, 177.

33. New England Wireless and Steam Museum, profile of George H. Corliss, http://www.newsm.org/steam-engines/george_h_corliss.html.

34. Morris, op. cit., 231.

35. Steven Ujifusa, "The Corliss Engine," The Philly History Blog, May 18, 2010, http://www.phillyhistory.org/blog/index.php/2010/05/the-corliss-engine/.

36. Quoted by Steven Ujifusa, "The Corliss Engine."

37. Morris, op. cit., 232.

38. Georg Auer, "A Genius Whose Three-wheeler Is Seen as the First Car,"

Autonews.com, undated, http://www.autonews.com/files/euroauto/inductees/benz.html.

39. Vaclav Smil, *Prime Movers of Globalization: The History and Impact of Diesel Engines and Gas Turbines* (Cambridge, MA: MIT Press, 2010), 29–30.

40. Roger Lowenstein, "The Crank That Set the World Rolling," *Wall Street Journal*, May 11, 2013, C6, http://online.wsj.com/article/SB10001424127887324266904578456621170724926.html.

41. James E. McClellan III and Harold Dorn, *Science and Technology in World History: An Introduction* (Baltimore: Johns Hopkins University Press, 2006), 340.

42. National Museum of the US Air Force data, http://www.nationalmuseum.af.mil/factsheets/factsheet.asp?id=513.

43. Some steam-powered trains went faster than 60 mph. But for many rail operators, 60 mph was a practical limit. For more, see Mike's Railway History, "Locomotive Speed Records," undated, http://mikes.railhistory.railfan.net/r070.html.

44. F1technical.net, "Formula One Engines," undated, http://www.f1technical.net/articles/4; Formula1.com, "Formula One Fuel and Oil—F1's Top Secret Tuning Aids," undated, http://www.formula1.com/news/features/2011/9/12524.html.

45. Author interview with Huibregtse by the author.

46. Google Finance data, http://www.google.com/finance?q=f&ed=us&ei=W-UJUYiMCZ6-lgOUzwE.

47. Google Finance data, http://www.google.com/finance?q=NYSE%3AF&fstype=ii&ed=us&ei=g-UJUbC9EJGClgOLGA.

48. Inflation-adjusted values calculated using The Inflation Calculator at http://www.westegg.com/inflation/infl.cgi.

49. Wikipedia.org, http://en.wikipedia.org/wiki/File:Henry_Ford_and_Barney_Oldfield_with_Old_999,_1902.jpg.

50. For current Ford prices, see cars.com.

51. Wikipedia.org, http://en.wikipedia.org/wiki/Ford_Model_T.

52. Ford.com, http://www.ford.com/cars/fiesta/specifications/.

53. Ford.com for the Fiesta specs: http://www.ford.com/cars/fiesta/specifications/. For the Model T specs, see: http://www.mtfca.com/discus/messages/29/34854.html?1188329164.

54. Max Giles, "Ford's 1.0-liter EcoBoost Wins Engine of the Year," Autoweek.com, June 5, 2013, http://www.autoweek.com/article/20130605/CARNEWS/130609916.

55. Here's the math: 22 hp in the Model T divided by 2.9 l = 7.6 hp/l. The Focus engine produces 123 hp/l. Thus, 118 / 7.6 = 15.5. For the Model T efficiency, see Ford.com, http://media.ford.com/article_display.cfm?article_id=858, which says the first Model T got between 13 and 21 mpg. For the EcoBoost, combined mileage is 32 to 41 mpg. See: Ray Hutton, "2012 Ford Focus

1.0L EcoBoost," *Car and Driver*, May 2012, http://www.caranddriver.com/news /2012-ford-focus-10l-ecoboost-first-drive-review.

56. Ford.com, http://corporate.ford.com/news-center/press-releases-detail/pr -ford26rsquos-10liter-ecoboost-wins-36658.

57. Http://www.performanceoiltechnology.com/syntheticoilandturbochargers .html.

58. Ray Hutton, "2012 Ford Focus 1.0L EcoBoost."

59. Ford.com, http://www.at.ford.com/news/cn/Pages/Ford%20Focus%20 1–0-litre%20EcoBoost%20Sets%2016%20World%20Speed%20Records.aspx.

Notes to Chapter 9

1. Apple press release, March 28, 2011, http://www.apple.com/pr/library/2011 /03/28Apple-Worldwide-Developers-Conference-to-Kick-Off-June-6-at -Moscone-West-in-San-Francisco.html.

2. David Gardner, "Gaunt and frail, cancer battle takes its toll on Steve Jobs in first picture since he left Apple," *Daily Mail*, August 28, 2011, http://www.daily mail.co.uk/news/article-2031100/Gaunt-frail-cancer-battle-takes-toll-Steve -Jobs-picture-left-Apple.html.

3. Thenextweb.com, "Steve Jobs' Last Public Appearance," October 7, 2011, http:// thenextweb.com/apple/2011/10/07/steve-jobs-last-public-appearance-video/.

4. Rich Miller, "Steve Jobs Provides a Look Inside the iDataCenter," Data Center Knowledge, June 6, 2011, http://www.datacenterknowledge.com/archives /2011/06/06/steve-jobs-provides-a-look-inside-the-idatacenter/. For size of Walmart stores, see: Reuters, http://www.reuters.com/finance/stocks/company Profile?symbol=WMT.N.

5. CloudTweaks.com, http://www.cloudtweaks.com/2012/05/review-of-the -top-cloud-storage-providers-in-the-consumer-segment/. See also: http://windows .microsoft.com/en-HK/skydrive/any-file-anywhere.

6. The 500,000 photo estimate comes from Flickr, based on 6.5 megapixel photos. See: flickr.com.

7. mkomo.com, "A History of Storage Cost," undated, http://www.mkomo .com/cost-per-gigabyte.

8. Holman W. Jenkins Jr., "Will Google's Ray Kurzweil Live Forever?" *Wall Street Journal*, April 12, 2013, http://online.wsj.com/article/SB1000142412788732 4504704578412581386515510.html.

9. John Gantz and David Reinsel, "Extracting Value from Chaos," IDC, June 2011, http://www.emc.com/collateral/analyst-reports/idc-extracting-value-from -chaos-ar.pdf.

10. A 2011 analysis by the Library of Congress put the size of the print collection alone at 208 terabytes. See: Mike Ashenfelder, "Transferring 'Library of Congress' of Data," July 11, 2011, loc.gov, http://blogs.loc.gov/digitalpreservation

/2011/07/transferring-libraries-of-congress-of-data/. For data on the LoC, see: http://www.loc.gov/about/facts.html.

11. To be clear on the math: The LoC collection is 200 terabytes or $2x10^{14}$. And 8 zettabytes is $8x10^{21}$. Divide the former into the latter and you get $4x10^{7}$, or 40 million.

12. Duncan Geere, "Google blimps will carry wireless signal across Africa," May 26, 2013, Wired.co.uk, http://www.wired.co.uk/news/archive/2013–05/26 /google-blimps/viewgallery/304505.

13. Intel, "Moore's Law and Intel Innovation," undated. http://www.intel.com /content/www/us/en/history/museum-gordon-moore-law.html.

14. Intel data on the company's 45 nanometer technology; http://download .intel.com/pressroom/kits/45nm/SandToCircuit_FINAL.pdf.

15. Http://en.wikipedia.org/wiki/M114_155_mm_howitzer.

16. Martin H. Weik, "The ENIAC Story," *Ordnance*, January–February 1961, http://ftp.arl.mil/~mike/comphist/eniac-story.html. Also, George Dyson, *Turing's Cathedral*, 22.

17. Data from photo of Eniac-on-a-chip supplied by Jan Van der Spiegel, University of Pennsylvania.

18. Jan Van der Spiegel, "ENIAC-on-a-chip," Penn Printout, March 1996, http://www.upenn.edu/computing/printout/archive/v12/4/chip.html.

19. One square meter contains 1 million square millimeters. Therefore, ENIAC covered 22,300,000 square millimeters. Divide that number by 40 and you get 557,500.

20. The average postage stamp has an area of about 400 square millimeters. See: http://www.bluebulbprojects.com/measureofthings/doShowResults.asp ?comp=area&unit=mm2&amt=5600&sort=pr&p=1.

21. Nobelprize.org, http://www.nobelprize.org/nobel_prizes/physics/laureates /1967/bethe-lecture.html. Source of quote: http://en.wikipedia.org/wiki/John_von _Neumann#cite_note-Life_Magazine_1957.2C,89, 104–44.

22. John R. Edwards, "A History of Early Computing at Princeton," undated, Princeton University, http://www.princeton.edu/turing/alan/history-of -computing-at-p/.

23. George Dyson, *Turing's Cathedral: The Origins of the Digital Universe* (New York: Pantheon, 2012), 59.

24. Ibid., 60.

25. Ibid., x.

26. Peter Huber and Mark P. Mills, *The Bottomless Well: The Twilight of Fuel, the Virtue of Waste, and Why We Will Never Run Out of Energy* (New York: Basic Books, 2005), 34, 35. See also, Tao Pang, *An Introduction to Computational Physics* (Cambridge: Cambridge University Press, 2006), 6, http://www.fisica.ufmg .br/~dickman/transfers/comp/pang06.pdf.

27. Van der Spiegel, "ENIAC-on-a-chip."

28. My home is about 2,700 square feet. I did not count the power density to include natural gas, as the heating load for my home is relatively small. The cooling load, given the hot summers in Texas, is relatively large.

29. Jesse H. Ausubel, "Renewable and Nuclear Heresies," *International Journal of Nuclear Governance, Economy and Ecology* 1, no. 3 (2007): 231, http://phe.rockefeller.edu/docs/HeresiesFinal.pdf.

30. Benj Edwards, "Birth of a Standard: The Intel 8086 Microprocessor," *PC World*, June 17, 2008, http://www.pcworld.com/article/146957/birth_of_a_standard_the_intel_8086_microprocessor.html.

31. Intel.com, http://www.intel.com/pressroom/kits/quickreffam.html.

32. Wikipedia, http://en.wikipedia.org/wiki/List_of_Intel_microprocessors; the chip referenced here is the Core i7 Ivy Bridge.

33. Wikipedia, http://en.wikipedia.org/wiki/List_of_Intel_microprocessors: the chip referenced here is the Core i7 Sandy Bridge-E, which has 8 cores and 2.27 billion transistors.

34. Pang, *An Introduction to Computational Physics*, 6, http://www.fisica.ufmg.br/~dickman/transfers/comp/pang06.pdf.

35. Dyson, op. cit., 75.

36. Dyson, op. cit., 85.

37. Dyson, op. cit., ix.

38. John R. Edwards, "A History of Early Computing at Princeton," undated, Princeton University, http://www.princeton.edu/turing/alan/history-of-computing-at-p/.

39. Dyson, op. cit., 6.

40. VPRO Backlight, "George Dyson on MANIAC, the First RAM," You Tube, January 27, 2011, http://www.youtube.com/watch?v=tWY5I2BvmH0.

41. Dyson, op. cit., 149.

42. http://www.washingtonpost.com/business/facebook-joins-ranks-of-largest-ipos-in-us-history/2012/05/18/gIQABxtuYU_gallery.html#photo=8.

43. Greenpeace, "How Clean Is Your Cloud?" April 2012, http://www.greenpeace.org/international/Global/international/publications/climate/2012/iCoal/HowCleanisYourCloud.pdf, 7.

44. Greenpeace.org, http://www.greenpeace.org/international/en/campaigns/climate-change/cool-it/ITs-carbon-footprint/Facebook/.

45. James Hamilton, "I Love Solar Power But . . . ," March 17, 2012, http://perspectives.mvdirona.com/2012/03/17/ilovesolarpowerbut.aspx.

46. Rich Miller, "Another Major Data Center for Prineville?," *Data Center Knowledge*, April 9, 2012, http://www.datacenterknowledge.com/archives/2012/04/09/another-major-data-center-for-prineville/.

47. Amanda S. Adams and David W. Keith, "Are Global Wind Power Resource Estimates Overstated?" *Environmental Research Letters*, February 25, 2013, http://iopscience.iop.org/1748-9326/8/1/015021/pdf/1748-9326_8_1_015021.pdf.

48. Central park is 843 acres, or 3.4 square kilometers: http://www.centralpark nyc.org/visit/general-info/faq/.

49. Jonathan Koomey estimates US data centers use 2 percent of domestic power. See: http://www.koomey.com/post/8323374335. In 2010, US power generation was 4326 terawatt-hours. See BP Statistical Review.Czech population data is here: http://www.google.com/publicdata/explore?ds=d5bncppjof8f9_&met _y=sp_pop_totl&idim=country:CZE&dl=en&hl=en&q=czech+republic+pop ulation.

50. According to the BP Statistical Review of World Energy 2012, in 2011, US solar energy production was 1.8 terawatt-hours.

51. Jonathan Koomey, "My New Study of Data Center Electricity Use in 2010," Koomey.com, July 31, 2011, http://www.koomey.com/post/8323374335.

52. BP Statistical Review of World Energy 2011.

53. Mark Mills, "The Cloud Begins with Coal," Digital Power Group, July 2013, http://www.tech-pundit.com/wp-content/uploads/2013/07/Cloud_Begins _With_Coal.pdf?c761ac&c761ac, 3.

54. Jonathan Koomey, "My New Study of Data Center Electricity Use in 2010."

55. BP Statistical Review of World Energy 2011. Between 2005 and 2010, global electricity use grew by 16.7%.

56. Adrian Bridgwater, "Intel 15 Billion Online Toasters by 2015," Computer weekly.com, September 14, 2012, http://www.computerweekly.com/cgi-bin /mt-search.cgi?blog_id=113&tag=Internet%20of%20Things&limit=20.

57. Greg Meckbach, "Ericsson Predicts 50 Billion Networked Devices," IT World Canada, February 26, 2010, http://www.itworldcanada.com/news/ericsson -predicts-50-billion-networked-devices/140089.

68. Current as of October 17, 2013, http://finance.yahoo.com/q?s=INTC&ql=0.

59. Http://www.google.com/finance?q=NASDAQ%3AINTC&fstype=ii&ed =us&ei=pv4bUZj3KaTAlgOCMQ.

60. Intel.com, http://download.intel.com/newsroom/kits/22nm/pdfs/22nm _Fun_Facts.pdf.

61. Intel.com, http://download.intel.com/newsroom/kits/22nm/pdfs/22nm _Fun_Facts.pdf.

62. Sharon Gaudin, "The Transistor: The Most Important Invention of the 20th Century?" Computerworld.com, December 12, 2007, http://www .computerworld.com/s/article/9052781/The_transistor_The_most_important_ invention_of_the_20th_century.

63. PC Magazine Encyclopedia, http://www.pcmag.com/encyclopedia_term /0,1237,t=25+process&i=49759,00.asp.

64. Ibid.

65. Intel.com, http://download.intel.com/newsroom/kits/22nm/pdfs/22nm _Fun_Facts.pdf.

66. Diedtra Henderson, "Moore's Law Still Reigns: How Computers Became

Smaller, Faster, Cheaper," *Seattle Times*, November 24, 1996, http://community. seattletimes.nwsource.com/archive/?date=19961124&slug=2361376.

67. Robert P. Colwell, *The Pentium Chronicles: The People, Passion, and Politics Behind Intel's Landmark Chips* (Hoboken: John Wiley & Sons, 2006), 4.

68. Yahoo! Finance data, http://finance.yahoo.com/q?s=intc&ql=1.

69. Ivan Smith, "Cost of Hard Drive Storage Space," undated, http://ns1758 .ca/winch/winchest.html.

70. Mkomo.com, "A History of Storage Cost," undated, http://www.mkomo .com/cost-per-gigabyte.

Notes to Chapter 10

1. Philips.com, "Philips Celebrates 25th Anniversary of the Compact Disc," August 16, 2007, http://www.newscenter.philips.com/main/standard/about/news /press/20070816_25th_anniversary_cd.wpd.

2. My vinyl collection holds 2,490 songs and weighs 62.3 kilos. Getting to 40,000 songs would be 16 times as much as my LP collection, therefore 62.3 × 16 = 1,000 kilos or 2,200 pounds. For the Fiat's specs, see: http://www.media .chrysler.com/dcxms/assets/specs/2012_Fiat_500_Specs.pdf. The iPod Classic has a volume of 67 cubic centimeters. Multiply that by 20,000 and you get 1.34 million cc or about 79,330 cubic inches. The refrigerator: an upright refrigerator/freezer has a volume of 73,278 cubic inches. For iPod Classic specs, see: http://www.apple .com/ipodclassic/specs.html.

3. Library of Congress, "The History of the Edison Cylinder Phonograph," undated, http://memory.loc.gov/ammem/edhtml/edcyldr.html.

4. Measurements based on author's personal collection, normalizing a 250-album collection of vinyl versus compact disc.

Notes to Chapter 11

1. Odyssey Marine Exploration, "SS *Gairsoppa* Historical Overview," undated, http://www.shipwreck.net/ssgairsoppahistoricaloverview.php.

2. Michael Winter, "$230M in silver found in WWII shipwreck off Ireland," *USA Today*, September 26, 2011, http://content.usatoday.com/communities/ondeadline /post/2011/09/230m-in-silver-found-in-ww2-shipwreck-off-ireland/1.

3. Huffington Post, "SS *Gairsoppa* Shipwreck Nets $38 Million in Silver," July 18, 2012, http://www.huffingtonpost.com/2012/07/18/ss-gairsoppa-shipwreck -38-million-silver_n_1683505.html. For more on Odyssey, see: http://finance. yahoo.com/q?s=OMEX.

4. Discovery.com, "Record Treasure Hauled from Shipwreck," July 18, 2012, http://news.discovery.com/history/biggest-treasure-yet-hauled-from-shipwreck -120718.htm#mkcpgn=fbdsc8.

5. Search results from BibleGateway.com. See: http://www.biblegateway.com/keyword/?search=gold&version1=NIV&searchtype=all.

6. David Wolman, *The End of Money: Counterfeiters, Preachers, Techies, Dreamers—And the Coming Cashless Society*, (Boston: Da Capo Press, 2012), 14.

7. James Surowiecki, "A Brief History of Money," *IEEE Spectrum*, June 2012, http://spectrum.ieee.org/at-work/innovation/a-brief-history-of-money/0.

8. Ibid.

9. World Bank data, http://databank.worldbank.org/databank/download/GDP.pdf.

10. FederalReserve.gov data, http://www.federalreserve.gov/paymentsystems/fedfunds_ann.html.

11. Toby Shapshak, "Long May the SMS Reign," TimesLive.co.za, June 11, 2012, http://www.timeslive.co.za/thetimes/2012/06/11/long-may-the-sms-reign.

12. Robert Bryce, "Printing Money Proves a Lucrative Trade," *Christian Science Monitor*, December 21, 1992, http://www.csmonitor.com/1992/1221/21072.html.

13. For more, see CSAnotes.com.

14. Jacob Goldstein and David Kestenbaum, "Why We Left the Gold Standard," NPR.com, April 21, 2011, http://www.npr.org/blogs/money/2011/04/27/135604828/why-we-left-the-gold-standard.

15. *The Week*, "Why Did the U.S. Abandon the Gold Standard?" October 5, 2012, http://mentalfloss.com/article/12715/why-did-us-abandon-gold-standard.

16. Donald L. Barlett and James B. Steele, "Billions over Baghdad," *Vanity Fair*, October 2007, http://www.vanityfair.com/politics/features/2007/10/iraq_billions200710.

17. David Barboza, "Chinese Way of Doing Business: In Cash We Trust, Exclusively," *New York Times*, May 1, 2013, http://www.nytimes.com/2013/05/01/business/global/chinese-way-of-doing-business-in-cash-we-trust.html.

18. World Bank data, via google.com/publicdata.

19. For more on M-PESA, see: http://www.safaricom.co.ke/personal/m-pesa/m-pesa-resource-centre/presentations.

20. That's the weight of a Nokia 3120b, as weighed by the author.

21. Miguel Helft, "The Death of Cash," *Fortune*, July 9, 2012, http://tech.fortune.cnn.com/2012/07/09/dorsey-square-death-cash/.

22. Rimma Kats, "Starbucks Generates More Than 3M Mobile Payment Transactions Per Week," MobileCommerceDaily.com, March 22, 2013, http://www.mobilecommercedaily.com/starbucks-generates-more-than-3m-mobile-payment-transactions-per-week.

23. Gartner.com, "Gartner Says Worldwide Mobile Payment Transaction Value to Surpass $171.5 Billion," May 29, 2012, http://www.gartner.com/newsroom/id/2028315.

24. Data from mobiletransaction.org, http://www.mobiletransaction.org/growing-sms-payments-world/.

25. For an analysis of the countries most ready to join the mobile payments sector, look at MasterCard's Mobile Payments Readiness Index: http://mobilereadiness.mastercard.com/the-index/.

26. Toby Shapshak statement at SXSW Interactive, Austin, TX, March 9, 2012. See: http://storify.com/softwaremono/the-dollar-100bn-mobile-bullet-train-called-africa.

27. Figure current as of October 10, 2013. Stock quote available via Reuters: http://www.reuters.com/finance/stocks/overview?symbol=SCOM.NR On October 10, 2013, Safaricom's market cap was 366B Kenyan shillings.

28. In fiscal year 2012, Safaricom had revenue of 107 billion Kenyan shillings. See the company's annual report for the year ended 31ˢᵗ March 2012, http://www.safaricom.co.ke/safaricom_annual_report/pdfs/Safaricom_Annual_Report.pdf, 45.

29. *The Economist,* "Is It a Phone, Is It a Bank?" March 20, 2013, http://www.economist.com/news/finance-and-economics/21574520-safaricom-widens-its-banking-services-payments-savings-and-loans-it.

30. Safaricom data, "M-PESA—An Overview," 2008, http://www.safaricom.co.ke/images/Downloads/Personal/M-PESA/Presentations/2008.09.01-m-pesa_media_workshop.pdf, 3, 4.

31. Safaricom annual report, March 2012, http://investinginafrica.net/wp-content/uploads/2012/08/Safaricom_Annual_Report_2012.pdf, 2.

32. Jerry Brito, "A Shift Toward Digital Currency," *New York Times*, October 16, 2012, http://www.nytimes.com/roomfordebate/2012/04/04/bringing-dollars-and-cents-into-this-century/a-shift-toward-digital-currency.

33. Nicole Perlroth, "Anonymous Payment Schemes Thriving on Web," May 29, 2013, *New York Times,* http://www.nytimes.com/2013/05/30/technology/anonymous-payment-schemes-thriving-on-web.html?_r=0.

34. Jessica Leber, "Cashing Out of Corruption," *MIT Technology Review,* March 19, 2012, http://www.technologyreview.com/news/427267/cashing-out-of-corruption/.

35. Nathan Hodge, "As U.S. Departs, Afghan Business Dries Up," *Wall Street Journal,* April 4, 2013, http://online.wsj.com/news/articles/SB10001424127887324281004578356240553085684.

36. Jessica Leber, "Cashing Out of Corruption."

37. Richard Shaffer, "Mobile Payments Gain Traction Among India's Poor," *New York Times*, December 4, 2013, http://www.nytimes.com/2013/12/05/business/international/mobile-payments-gain-traction-among-indias-poor.html?hpw&rref=business.

38. Tom Chatfield, "Bitcoin and the Illusion of Money," BBC.com, April 12, 2013, http://www.bbc.com/future/story/20130412-bitcoin-and-the-illusion-of-money.

39. See: https://venmo.com/.

Notes to Chapter 12

1. For more on this, see: Edward O. Wilson, "The Riddle of the Human Species," *New York Times,* February 24, 2013, http://opinionator.blogs.nytimes.com/2013/02/24/the-riddle-of-the-human-species/?_r=0.

2. Joel Kotkin, *The City: A Global History,* (New York: The Modern Library, 2005), 4.

3. Worldatlas.com, http://www.worldatlas.com/citypops.htm#.UauR0yt4aRk.

4. UN data, http://www.un.org/esa/population/publications/longrange2/WorldPop2300final.pdf.

5. Edward Glaeser, *Triumph of the City: How Our Greatest Invention Makes Us Richer, Smarter, Greener, Healthier and Happier* (New York: Penguin Press, 2011) 1.

6. Steven Johnson, *Where Good Ideas Come From: The Natural History of Innovation* (New York: Riverhead Books, 2010), 10.

7. Gerald Carlino, Satyajit Chatterjee, and Robert Hunt, "Matching and Learning in Cities: Urban Density and the Rate of Invention," Federal Reserve Bank of Philadelphia, April 2005, http://www.philadelphiafed.org/research-and-data/publications/working-papers/2004/wp04–16.pdf, ii.

8. See: http://cybersocialstructure.org/category/cybersocialstructure/.

9. Stewart Brand, *Whole Earth Discipline: An Ecopragmatist Manifesto* (New York: Viking, 2009), 26.

10. Kai Ryssdal, "A Pragmatic Response to Climate Change," Marketplace, October 26, 2009, http://www.marketplace.org/topics/business/pragmatic-response-climate-change.

11. Richard Dobbs, Jaana Remes, James Manyika, Charles Roxburgh, Sven Smit and Fabian Schaer, "Urban World: Cities and the Rise of the Consuming Class," McKinsey Global Institute, June 2012, http://www.mckinsey.com/insights/mgi/research/urbanization/urban_world_cities_and_the_rise_of_the_consuming_class.

12. Ibid., 1.

13. Ibid., 4.

14. Bruce Katz and Jennifer Bradley, "Metro Connection," *Democracy,* Spring 2011, http://www.democracyjournal.org/20/metro-connection.php?page=all.

15. Kotkin, *The City,* op. cit., xx.

Notes to Chapter 13

1. Cited by Keith Fuglie, "Why the Pessimists Are Wrong," *IEEE Spectrum,* June 2013, http://spectrum.ieee.org/green-tech/conservation/agricultural-productivity-will-rise-to-the-challenge.

2. Geohive.com data, http://www.geohive.com/earth/his_history3.aspx.

3. For the text of the book, see http://www.constitution.org/cmt/malthus /population.html.

4. Fuglie, op. cit.

5. Market cap data as of October 17, 2013. See: http://www.google.com/finance ?cid=656159.

6. Organic Trade Association, "Industry Statistics and Projected Growth," un-dated, http://www.ota.com/organic/mt/business.html.

7. Verena Seufert, Navin Ramankutty, and Jonathan Foley, "Comparing the Yields of Organic and Conventional Agriculture," *Nature*, 485, 229–232, May 10, 2012, http://www.nature.com/nature/journal/v485/n7397/full/nature11069.html.

8. James E. McWilliams, "Organic Crops Alone Can't Feed the World," *Slate*, March 10, 2011, http://www.slate.com/articles/health_and_science/green_room /2011/03/organic_crops_alone_cant_feed_the_world.html.

9. Steve Savage, "Today's Organic, Yesterday's Yields," Biofortified, February 10, 2011, http://www.biofortified.org/2011/02/todays-organic-yesterdays-yields/.

10. Earth Policy Institute data, http://www.earth-policy.org/data_center/C24.

11. International Food Policy Research Institute, "Investing in Agriculture to Reduce Poverty and Hunger," 2012, http://www.ifpri.org/publication/investing -agriculture-reduce-poverty-and-hunger.

12. Earth Policy Institute data, "World Grain Production, Area, and Yield, 1950–2011," http://www.earth-policy.org/data_center/C24/.

13. Greenpeace data, http://www.greenpeace.org/international/en/campaigns /agriculture/problem/genetic-engineering/.

14. *The Scientist*, "AAAS: Don't Label GM Foods," October 30, 2012, http://www.the -scientist.com/?articles.view/articleNo/33057/title/AAAS—Don-t-Label -GM-Foods/.

15. Henry Miller, "Gold Rice, Red Tape," *Guardian*, October 17, 2008, http://www.guardian.co.uk/commentisfree/cifamerica/2008/oct/17/gm-crops -golden-rice.

16. WHO data, http://www.who.int/nutrition/topics/vad/en/.

17. Greenpeace.org, "Golden Illusion: The Broken Promises of 'Golden' Rice," September 2012, http://www.greenpeace.org/seasia/ph/PageFiles/462570 /Golden%20Illusion.pdf.

18. See: http://www.greenpeace.org/usa/en/campaigns/.

19. Dan Charles, "In a Grain of Golden Rice, A World of Controversy over GMO Foods," NPR.org, March 7, 2013, http://www.npr.org/blogs/the salt/2013/03/07/173611461/in-a-grain-of-golden-rice-a-world-of-controversy -over-gmo-foods.

20. Jesse H. Ausubel, Iddo K. Wernick, and Paul E. Waggoner, "Peak Farm-land and the Prospect for Land Sparing," Rockefeller University Program for the Human Environment, 2012, http://phe.rockefeller.edu/docs/PDR.SUPP%20 Final%20Paper.pdf, 19.

Notes to Chapter 14

1. See: http://www.netindex.com/download/allcountries/.

2. CIA World Factbook, https://www.cia.gov/library/publications/the-world
-factbook/rankorder/2004rank.html.

3. James Fallows, "Can China Escape the Low-Wage Trap?" *New York Times*,
May 27, 2012, http://www.nytimes.com/2012/05/27/opinion/sunday/can-china
-escape-the-low-wage-trap.html?pagewanted=all.

4. Economist Intelligence Unit, "Democracy Index 2011: Democracy Under
Stress," December 2011, http://www.eiu.com/Handlers/WhitepaperHandler
.ashx?fi=Democracy_Index_Final_Dec_2011.pdf&mode=wp&campaignid=
DemocracyIndex2011, 3–8.

5. Committee to Protect Journalists, "10 Most Censored Countries," May 2,
2012, http://www.cpj.org/reports/2012/05/10-most-censored-countries.php.

6. Alexander Martin, "Mobile Phones Proliferate in North Korea," *Wall Street
Journal*, July 28, 2012, http://online.wsj.com/news/articles/SB1000142405270230
44586045774875721176127932.

7. CIA World Factbook, https://www.cia.gov/library/publications/the-world
-factbook/fields/2119.html.

8. See: http://www.indexmundi.com/world/gdp_per_capita_(ppp).html.

9. Marc Frank, "More Cubans Have Local Intranet, Mobile Phones," Reuters, June 15, 2012, http://www.reuters.com/article/2012/06/15/net-us-cuba
-telecommunications-idUSBRE85D14H20120615.

10. Committee to Protect Journalists, http://www.cpj.org/reports/2012/05
/10-most-censored-countries.php.

11. Electronic Frontier Foundation, "This Week in Internet Censorship: India, Iran, Brazil, Russia, and More," May 15, 2012, https://www.eff.org/deep
links/2012/05/week-internet-censorship-iran-censors-internet-censorship
-decree-indian-government.

12. ONI is a collaborative partnership of three entities: the Citizen Lab at the
Munk School of Global Affairs at the University of Toronto; the SecDev Group,
an Ottawa-based consulting firm; and the Berkman Center for Internet and Society at Harvard University.

13. CIA World Factbook, https://www.cia.gov/library/publications/the-world
-factbook/rankorder/2004rank.html.

14. Https://www.cia.gov/library/publications/the-world-factbook/rankorder
/2004rank.html.

15. World Bank press release, "Maximizing Mobile," July 17, 2012.

16. Andrew Pollack, "City's Mobile Phone Battle," *New York Times*, March 3,
1984, 29.

17. Associated Press, "First Cell Phone a True 'Brick,'" April 11, 2005, http://
www.msnbc.msn.com/id/7432915/ns/technology_and_science-wireless/t/first
-cell-phone-true-brick/#.UEV__Wie5M4.

18. Retrobrick.com data, http://www.retrobrick.com/moto8000.html.

19. Northern Voices Online, "Samsung Galaxy Note 2 Vs Galaxy S3," September 2, 2012, http://nvonews.com/2012/09/02/samsung-galaxy-note-2-vs-galaxy -s3-comparison-of-specs-features/.

20. Amazon data, http://www.amazon.com/Samsung-Galaxy-Unlocked-Smart -Marble/dp/B0080DJ6CM.

21. Jonathan Charles, "Samsung Galaxy S3 Gets Cheaper: Sub-$200 Price May Be Arriving," Mobilenapps.com, September 2, 2012, http://www.mobilen apps.com/articles/4034/20120902/samsung-galaxy-s3-gets-cheaper-sub-200 .html.

22. Http://smartphones.techcrunch.com/l/245/Samsung-Galaxy-S3.

Notes to Chapter 15

1. Matt Hinton, "Johnny Manziel, Big Man on Campus, Avoiding Campus in Web-only Classes," CBSsports.com, February 18, 2013, http://www .cbssports.com/collegefootball/blog/eye-on-college-football/21727707 /johnny-manziel-big-man-on-campus-avoiding-campus-in-web-only -classes.

2. Tamar Levin, "Instruction for Masses Knocks Down Campus Walls," *New York Times*, March 4, 2012, http://www.nytimes.com/2012/03/05/education/moocs -large-courses-open-to-all-topple-campus-walls.html?pagewanted=all&_r=0.

3. Http://www.nytimes.com/2011/08/16/science/16stanford.html?_r=1.

4. For more, see: http://clep.collegeboard.org.

5. Jeff Denneen and Tom Dretler, "The Financially Sustainable University," Bain & Company, July 6, 2012, http://www.bain.com/publications/articles /financially-sustainable-university.aspx.

6. Jasmine Evans, "The Rising Cost of Tuition Surpasses the Rate of Inflation," Diverseeducation.com, February 11, 2013, http://diverseeducation.com /article/51243/#.

7. Cameron McWhirter and Douglas Belkin, "Student Drought Hits Smaller Universities," *Wall Street Journal*, July 26, 2013, A3.

8. Mark Suster, "In 15 Years from Now Half of U.S. Universities May Be in Bankruptcy. My Surprise Discussion with @ClayChristensen," Bothsidesof thetable.com, March 3, 2013, http://www.bothsidesofthetable.com/2013/03/03 /in-15-years-from-now-half-of-us-universities-may-be-in-bankruptcy-my -surprise-discussion-with-claychristensen/.

9. Wikipedia.org, http://en.wikipedia.org/wiki/Jean-Baptiste_de_La_Salle.

10. Wikipedia.org, http://en.wikipedia.org/wiki/Salman_Khan_(educator).

11. Wikipedia.org, http://en.wikipedia.org/wiki/Salman_Khan_(educator).

12. Wikipedia.org, http://en.wikipedia.org/wiki/Khan_Academy.

13. Khan Academy data, http://www.khanacademy.org/about.

14. Scienceexchange.com, "MOOCs and Khan Academy: Opening up Science Education," December 12, 2012, http://blog.scienceexchange.com/2012/12/moocs-and-khan-academy-opening-up-science-education/.

15. A.J. Jacobs, "Two Cheers for Web U," *New York Times*, April 21, 2013, http://www.nytimes.com/2013/04/21/opinion/sunday/grading-the-mooc-university.html?src=recg.

16. Ibid.

17. David Brooks, "The Practical University," *New York Times*, April 4, 2013, http://www.nytimes.com/2013/04/05/opinion/Brooks-The-Practical-University.html?ref=todayspaper&_r=2&.

Notes to Chapter 16

1. Foxnews.com, "Real-life 'Star Trek' Tricorder Project Raises $1 Million," http://www.foxnews.com/science/2013/06/25/real-life-star-trek-tricorder-project-raises-1-million/#ixzz2ZQ1vPkJV.

2. Indiegogo.com data, http://www.indiegogo.com/projects/scanadu-scout-the-first-medical-tricorder?c=home.

3. Data collected and analyzed by author based on Kenneth J. Nazinitsky and Burton M. Gold, "Radiology—Then and Now," *American Journal of Roentgenology*, August 1988, http://www.ajronline.org/content/151/2/249.full.pdf, as well as inflation-adjusted cost data published by New Choice Health, http://www.newchoicehealth.com/X-Ray-Cost.

4. For more on this, see DMXworks.com.

5. Eliza Strickland, "The Gene Machine and Me," *IEEE Spectrum*, February 29, 2013, http://spectrum.ieee.org/biomedical/devices/the-gene-machine-and-me.

6. Amy Dockser Marcus, "The Future of Medicine is Now," *Wall Street Journal*, December 31, 2012, http://online.wsj.com/article/SB10001424127887323530404578205692226506324.html.

7. Nicholas Wade, "Decoding DNA with Semiconductors," *New York Times*, July 20, 2011, http://www.nytimes.com/2011/07/21/science/21genome.html.

8. American Diabetes Association data, http://www.diabetes.org/diabetes-basics/diabetes-statistics/.

9. Eric Topol, *The Creative Destruction of Medicine: How the Digital Revolution Will Create Better Health Care* (New York: Basic Books, 2012), 65–66. See also: http://www.medtronicdiabetes.com/products/guardiancgm.

10. Ben Coxworth, "Next-Generation Camera Pill Could Transmit HD Video from Inside the Body," Gizmag.com, November 28, 2011, http://www.gizmag.com/next-generation-camera-pill/20640/.

11. Joseph Walker, "PillCam Maker Given Imaging to Be Bought by Covidien," *Wall Street Journal*, December 8, 2013, http://online.wsj.com/news/articles/SB10001424052702304014504579246692001240318.

12. Camilla Andersson, "3-D Printer Redefines Reconstructive Surgery," *European Medical Device Technology,* January 15, 2013, http://www.emdt.co.uk /article/3-d-printer-redefines-reconstructive-surgery.

13. Jason Kane, "Health Costs: How the U.S. Compares with Other Countries," PBS NewsHour, October 22, 2012. In 2011, US GDP was $15 trillion.

Notes to Chapter 17

1. Cactus Drilling has a good primer that describes the jobs on the rig. See: http://cactusdrilling.com/careers.

2. The numbers vary every year, but in 2010, the *Daily Beast* reported that oil and gas drillers were experiencing 24 fatalities per 100,000 workers. Fishermen, loggers, and ranchers all had higher fatality rates. *Daily Beast*, "Dangerous Jobs," April 7, 2010, http://www.thedailybeast.com/galleries/2010/04/07/dangerous -jobs.html#slide14.

3. National Oilwell Varco, "Iron Roughnecks," undated, http://www.nov .com/Drilling/Iron_Roughnecks.aspx.

4. The operating cost for land rigs varies widely. But the rental rate for a top-drive rig is about $25,000 per day. When all other costs, including fuel, consulting firms, and other items, like drill bits, are added in, the total cost can reach $100,000 per day, or $4,166 per hour.

5. Transcript of State of the Union address, http://www.npr.org/2011/01/26 /133224933/transcript-obamas-state-of-union-address.

6. BP Statistical Review of World Energy 2012.

7. Bloomberg New Energy Finance, "New Investment in Clean Energy Fell 11% in 2012," January 14, 2013, http://about.bnef.com/2013/01/14/new-investment -in-clean-energy-fell-11-in-2012–2/.

8. IHS, "Total 2012 Upstream Oil and Gas Spending to Reach Record Level of Nearly $1.3 Trillion; Set to Exceed $1.6 Trillion by 2016, IHS Study Says," April 30, 2012, http://press.ihs.com/press-release/energy-power/total-2012-upstream -oil-and-gas-spending-reach-record-level-nearly-13-tri.

9. EIA data, http://www.eia.gov/dnav/ng/ng_enr_wellend_s1_a.html.

10. Ibid.

11. Guinness World Records, http://www.guinnessworldrecords.com/records -11000/largest-drillship/.

12. Personal communication by the author with Norman Naill, Devon's drilling superintendent, on January 23, 2013.

13. Southwestern Energy investor presentation, August 2013, http://www.swn .com/investors/LIP/latestinvestorpresentation.pdf.

14. EIA data, "Short-term Energy Outlook," February 14, 2013, http://www .eia.gov/forecasts/steo/special/pdf/2013_sp_02.pdf?scr=email.

15. Boston Company Asset Management, "End of an Era: The Death of Peak Oil,"

February 2013, http://www.thebostoncompany.com/assets/pdf/views-insights/Feb13_Death_of_Peak_Oil.pdf, 3.

16. In March 2013, Anadarko Petroleum Corp. and four partner companies announced another major discovery of oil in a formation known as the Lower Tertiary Trend. The well was drilled to a depth of 31,000 feet below the ocean floor in 5,800 feet of water. See: Dow Jones Newswires, "Anadarko, Partners Announce U.S. Gulf of Mexico Oil Find," March 19, 2013, http://m.foxbusiness.com/quickPage.html?content=91014937&page=32811&pageNum=-1. Tudor, Pickering, Holt & Co., a Houston-based investment banking firm estimates the Shenandoah-2 discovery has 3.7 billion barrels of oil equivalent. A few days after Anadarko announced its find, Chevron announced a find of similar magnitude in the Lower Tertiary. See: Zain Shauk, "Chevron Strikes Oil More Than 6 Miles Below Sea Level," FuelFix, March 25, 2013, http://fuelfix.com/blog/2013/03/25/chevron-finds-oil-more-than-6-miles-below-sea-level/?utm_source=twitterfeed&utm_medium=twitter.

17. Boston Company Asset Management, "End of an Era: The Death of Peak Oil," 3.

18. BP Statistical Review of World Energy 2012.

19. Abbott Payson Usher, *A History of Mechanical Inventions* (New York: Dover, 1982), 9.

20. Richard Heinberg, *The Party's Over: Oil, Water and the Fate of Industrial Societies,* (Gabriola Island, BC, Canada: New Society, 2003), 105.

21. *Los Angeles Times*, "U.S. Warned Oil Shortage Due in 20 Years," August 18, 1946, 6.

22. Heinberg, *The Party's Over*, 106.

23. Robert L. Bradley Jr., and Richard W. Fulmer, *Energy: The Master Resource* (Dubuque: Kendall/Hunt Publishing, 2004), 81.

24. Quoted in ibid, 81.

25. Ibid.

26. Matt Ridley, "Apocalypse Not: Here's Why You Shouldn't Worry About End Times," *Wired*, August 17, 2012, http://www.wired.com/wiredscience/2012/08/ff_apocalypsenot/all.

27. Ibid.

28. Robert A. Hefner, *The GET: The Grand Energy Transition* (Oklahoma City, GHK Exploration LLC, 2008), 35.

29. Mike Ruppert, "Interview with Matthew Simmons," August 18, 2003, http://www.oilcrash.com/articles/blackout.html.

30. Reuters, "Exxon Says N. America Gas Production Has Peaked," June 21, 2005, http://www.reuters.com/article/Utilities/idUSN2163310420050621.

31. *Oil and Gas Journal,* "Marketed Natural Gas Production," January 7, 2013, 34.

Notes to Chapter 18

1. CBS News, "Transcript of Barack Obama's Speech," February 10, 2007, http://www.cbsnews.com/stories/2007/02/10/politics/main2458099.shtml.

2. Http://fayobserver.com/articles/2013/03/02/1240799.

3. World Nuclear Association, "Heat Values of Various Fuels, March 5, 2010, http://world-nuclear.org/info/Facts-and-Figures/Heat-values-of-various-fuels/#.Ud3sIz54ZDs.

4. Wikipedia, http://en.wikipedia.org/wiki/Boeing_737.

5. See b737.org, http://www.b737.org.uk/techspecsdetailed.html.

6. Http://www.climate-one.org/transcripts/robert-f-kennedy-jr-transcript.

7. Bart Jansen, "Questions focus on Boeing's 787 Dreamliner's Batteries," *USA Today*, January 17, 2013, http://www.usatoday.com/story/travel/flights/2013/01/17/dreamliner-batteries/1842871/.

8. World Nuclear Association, "Heat values of various fuels."

9. A 40-foot shipping container covers 320 square feet or nearly 30 square meters. See: http://www.sjonescontainers.co.uk/container/dimensions.asp. If we assume the footprint of Clean Energy Systems' generator, and all of the related equipment, is 20 of the standard 40-foot shipping containers, then the entire site would cover 600 square meters. Therefore, with a power output of 70,000,000 watts / 600 m^2, the power density would be nearly 117,000 W/m^2.

Notes to Chapter 19

1. According to Peabody officials (Beth Sutton e-mail dated April 11, 2012), the mine's coal production can produce 205 terawatt-hours of electricity per year. In 2012, Mexico's electricity generation totaled 291 TWh. Source: BP Statistical Review of World Energy 2013.

2. EIA data, http://www.eia.gov/coal/annual/pdf/table1.pdf.

3. James Hansen, "Coal-fired Power Stations Are Death Factories. Close them," *The Observer*, February 14, 2009, http://www.guardian.co.uk/commentisfree/2009/feb/15/james-hansen-power-plants-coal.

4. EIA data, http://www.eia.gov/forecasts/ieo/coal.cfm.

5. BP Statistical Review of World Energy 2013. Between 2002 and 2012, global coal use jumped by 26.5 million barrels of oil equivalent per day. Over that same time frame, oil consumption grew by 11.3 million barrels of oil equivalent per day; natural gas increased by 14.3 million barrels of oil equivalent per day; and nuclear fell by 1 million barrels of oil equivalent per day, largely due to Japan's shutdown of its nuclear facilities after the Fukushima disaster. (Hydropower increased by 4.7 million barrels of oil equivalent per day. Wind increased by 2.1 million barrels of oil equivalent per day.)

6. Natural-gas.au, "Common Properties of Commercial Fuels," http://www.natural-gas.com.au/about/references.html.

7. IEA Key World Energy Statistics 2013, 24, http://www.iea.org/publications /freepublications/publication/name,31287,en.html.

8. Matt Ridley, *Rational Optimist: How Prosperity Evolves* (New York: Harper Collins, 2010), 223.

9. Barbara Freese, *Coal: A Human History* (New York: Penguin, 2003), 97.

10. Keith Bradsher and David Barboza, "Pollution from Chinese Coal Casts a Global Shadow," *New York Times*, June 11, 2006, http://www.nytimes.com/2006 /06/11/business/worldbusiness/11chinacoal.html?pagewanted=print.

11. Mine Safety and Health Administration data, http://www.msha.gov/stats /charts/coal2011yearend.asp.

12. John Raby, "Gary May, Former Upper Big Branch Mine Superintendent, Pleads Guilty to Federal Conspiracy Charges," *Huffington Post*, March 29, 2012, http://www .huffingtonpost.com/2012/03/29/gary-may-upper-big-branch_n_1387689.html.

13. Http://www.sfgate.com/business/bloomberg/article/Worst-India-Outage -Highlights-60-Years-of-Missed-3755620.php.

14. BP Statistical Review of World Energy 2013.

15. IEA Key World Energy Statistics 2013, 53.

16. Ibid., 51, 57. The per-capita electricity use in the US is 13,277 kWh per year.

17. Victor Mallet, "Indian Power Shortage Is Achilles Heel of Economy," *Financial Times,* May 30, 2013, http://www.ft.com/intl/cms/s/0/f5bc2d72-c8f1–11e2 –9d2a-00144feab7de.html#axzz2VABnbzCn.

18. Sanjeev Choudhary and John Chalmers, "RPT-India's Blackouts Shine Light on Broken Power Sector, Reuters, August 1, 2012, http://in.reuters.com /article/2012/08/01/india-power-idINL4E8J12R920120801.

19. Prashant Mehra and Abhishek Vishnoi, "Coal India Shares Slip on Import Risks; UK Fund Files Suit," August 1, 2012, http://in.reuters.com/article /2012/08/01/tci-coalindia-idINL4E8J12SD20120801.

20. *TheHindu.com.* "Pachauri Defends India's Climate Stand," July 22, 2009, http://www.hindu.com/thehindu/holnus/001200907220334.html.

21. IEA World Energy Outlook 2011, 381.

22. Http://www.mikebloomberg.com/index.cfm?objectid=4D1722F5-C29C -7CA2-FCB6385366A49867.

23. IEA data, http://www.worldenergyoutlook.org/resources/energydevelopment /accesstoelectricity/.

24. For electricity data: http://data.worldbank.org/indicator/EG.USE.ELEC. KH.

25. BP Statistical Review of World Energy 2011.

26. Ibid.

27. Todd Moss, "How Long Can You Live with This Kind of Modern Energy?" Center for Global Development, November 14, 2013, http://www.cgdev.org /blog/how-long-can-you-live-kind-%E2%80%9Cmodern%E2%80%9D-energy.

28. IEA, "Tracking Clean Energy Progress 2013," http://www.iea.org/publications /TCEP_web.pdf, 8.

29. World Nuclear Association, "Plans for New Reactors Worldwide," March 2013, http://www.world-nuclear.org/info/Current-and-Future-Generation/Plans -For-New-Reactors-Worldwide/#.UYI0FCt4aRk.

30. NETL data, http://www.netl.doe.gov/coal/refshelf/ncp.pdf, 16.

31. EIA data, http://www.eia.gov/electricity/annual/html/epa_04_03.html.

32. Dieter Helm, *The Carbon Crunch, How We're Getting Climate Change Wrong— and How to Fix It* (New Haven, CT: Yale University Press, 2012), 45.

33. Quirin Schiermeier, "Renewable Power: Germany's Energy Gamble," *Nature*, April 10, 2013, http://www.nature.com/news/renewable-power-germany-s -energy-gamble-1.12755.

34. Brad Stone, "A Subsidiary Charts Google's Next Frontier: Renewable Energy," *New York Times,* November 28, 2007, C3.

35. Google was incorporated in 1998. See: http://www.google.com/about /company/history/.

36. Google Inc., http://www.google.com/intl/en/press/pressrel/20071127_green .html.

37. Google data, http://googleblog.blogspot.com/2011/11/more-spring-cleaning -out-of-season.html.

38. Carnegie Mellon University, bio of Jay Whitacre, undated, http://www .materials.cmu.edu/people/whitacre.html.

39. J. F. Whitacre, T. Wiley, S. Shanbhag, S. Chun, W. Yang, D. Blackwood, A. Mohamed, E. Weber, D. Humphreys, "Large Format Aqueous Electroloyte Polyionic Devices for Low Cost, Multi-Hour Stationary Energy Storage," September 2012, Aquion Energy, http://cdn2.hubspot.net/hub/147472/file-23732056-pdf /docs/aquion_technical_presentation.pdf, 8.

40. Vaclav Smil, *Energy Myths and Realities: Bringing Science to the Energy Policy Debate (*Washington, DC: AEI Press, 2010), 19.

41. Freedoniagroup.com data, http://www.freedoniagroup.com/InTheNews Display.aspx?DocumentId=46888. See also, GIA report, August 29, 2011, which puts global consumer battery market at \$55.4 billion by 2017. See: http://www.prweb.com /releases/consumer_batteries/primary_secondary/prweb8605940.html.

42. Randy Alfred, "March 20, 1800: Volta's Battery Shows Potential," *Wired*, March 20, 2008, http://www.wired.com/science/discoveries/news/2008/03/dayin tech_0320.

43. Batteryuniversity.com data, http://batteryuniversity.com/learn/article/whats _the_best_battery.

44. Boeing data, http://www.boeing.com/787-media-resource/docs/fct -031514−787%20battery%20timeline.pdf. For grounding, see: CBSnews.com,

"FAA Grounds 787 Dreamliner," January 16, 2013, http://www.cbsnews.com/8301–201_162–57564384/faa-grounds-787-dreamliners/.

45. Karthikeyan Sundaram, "Air India Said to Seek $37 Million in Damages for 787 Grounding," *Bloomberg*, April 1, 2013, http://www.bloomberg.com/news/2013–04–01/air-india-said-to-seek-37-million-in-damages-for-787-grounding.html.

46. Clive Irving, "Boeing Won't Budge as Industry Abandons Lithium-Ion Battery," *Daily Beast*, April 25, 2013, http://www.thedailybeast.com/articles/2013/04/25/boeing-won-t-budge-as-industry-abandons-lithium-ion-battery.html.

47. CBSnews.com, "Another Sony Laptop Battery Recall," February 11, 2009, http://www.cbsnews.com/2100–500395_162–2051618.html.

48. Batteryuniversity.com data, http://batteryuniversity.com/learn/article/can_the_lead_acid_battery_compete_in_modern_times.

49. Benjamin Romano, "Fire Destroys 15MW Wind Energy Storage System in Hawaii," RechargeNews.com, August 8, 2012, http://www.rechargenews.com/business_area/innovation/article319427.ece.

50. Umair Irfan, "Battery Fires Reveal Risks of Storing Large Amounts of Energy," *Scientific American*, November 20, 2011, http://www.scientificamerican.com/article.cfm?id=battery-fires-risks-storing-lareg-amounts-energy.

51. Hydrogen has extremely high gravimetric energy density—141 million joules per kilogram, which is three times the energy density of diesel fuel. A mere spark can ignite hydrogen gas. (The most famous example of hydrogen's volatility is almost certainly the spectacular fire that engulfed the *Hindenburg* airship in 1937.) For comparison, high-quality coal has an energy density of 32 million joules per kilogram. And that coal requires a robust flame to be applied before it will be ignited. For data on energy density see: http://en.wikipedia.org/wiki/Energy_density.

52. Aquionenergy.com, http://www.aquionenergy.com/stationary-energy-storage-batteries.

53. For storage data, see Nissan, http://www.nissanusa.com/leaf-electric-car/versions-specifications?next=ev_micro.section_nav. For weight, see http://en.wikipedia.org/wiki/Nissan_Leaf.

54. Those numbers sound pretty good until you consider that gasoline contains about 12,000 watt-hours per kilogram. The high energy density of gasoline—about 150 times what's found in the battery used by the Leaf—explains why oil is an essential fuel for transportation. For the energy density of gasoline, see: MacGregor Campbell, "Energy on the Go," *New Scientist*, November 30, 2012, http://www.energyrealities.org/content/energy-on-the-go/erp7D81ABF7FC1E4B181.

55. KPCB.com data, http://www.kpcb.com/partner/al-gore.

56. Andrew Herndon, "Bill Gates Invests in Battery Maker Aquion Energy," *Bloomberg*, April 2, 2013, http://www.bloomberg.com/news/2013–04–02/bill-gates-invests-in-battery-maker-aquion-energy.html.

57. Http://www.greentechmedia.com/articles/read/aquion-energys-disruptive-battery-tech-picks-up-15m-loan.

58. A typical 12-volt car battery is rated at 85 amp-hours; 12 volts x 85 Ah = 1020 watt-hours. See: http://answers.yahoo.com/question/index?qid=20110116084042A Aoie9A. There are about 1 billion cars in the world. Therefore, 1 billion x 1020 watt-hours is roughly equal to 1 terawatt-hour. For auto numbers, see: wardsauto .com, http://wardsauto.com/ar/world_vehicle_population_110815.

59. BP Statistical Review of World Energy 2012.

60. At 22,500 TWh per year, we use 61 TWh per day, or about 2.5 TWh per hour. Thus, the world's auto batteries, with 1 TWh of storage, would only be able to provide less than 30 minutes of electricity for the world.

61. In 2012, total electricity sales were 22,500 terawatt hours, which is 22.5 trillion kilowatt-hours. Multiplied by $0.10 per kilowatt-hour (the approximate price of electricity in the US) equals $2.25 trillion.

Notes to Chapter 20

1. For height data, see these Wisconsin wind turbines, which are 500-feet tall. That's right at 150 meters. See: http://media.journalinteractive.com/images /BIGWIND10G.jpg.

2. Siemens.com data, http://www.swe.siemens.com/spain/web/es/energy /energias_renovables/eolica/Documents/6MW_direct_drive_offshore_wind_turbine .pdf.

3. Chris Callaway, "Dimensions of a UEFA Soccer Field," Livestrong.org, May 26, 2011, http://www.livestrong.com/article/384541-dimensions-of-a-uefa -soccer-field/.

4. K. Shawn Smallwood, "Comparing Bird and Bat Fatality-rate Estimates Among North American Wind-energy Projects," *Wildlife Society Bulletin*, March 26, 2013, http://onlinelibrary.wiley.com/doi/10.1002/wsb.260/abstract.

5. Http://www.windpoweringamerica.gov/images/windmaps/installed_capacity _2007.jpg.

6. Capacity figure here: http://www.windpoweringamerica.gov/images/windmaps /installed_capacity_2011.jpg.

7. This document—http://www.biologicaldiversity.org/campaigns/protecting _birds_of_prey_at_altamont_pass/pdfs/factsheet.pdf—puts the death toll at Altamont at 116 golden eagles per year. Golden Gate Audubon says it is up to 110. See: http://www.goldengateaudubon.org/conservation/birds-at-risk /avian-mortality-at-altamont-pass/.

8. Department of Justice, press release, http://www.justice.gov/opa/pr/2013/ November/13-enrd-1253.html.

9. RTE News, "Over 2,000 Protest over Wind Farms in the Midlands," July 27, 2013, http://www.rte.ie/news/player/2013/0727/3568500-over-2000

-protest-over-wind-farms-in-the-midlands/ For the number of turbines, see: http://www.irishtimes.com/news/environment/protesters-oppose-1–150-wind-farms-in-midlands-1.1377697.

10. Http://www.epaw.org/index.php?lang=en.

11. Http://www.countryguardian.net/WAG%20List.html.

12. Http://www.windaction.org/orglist.

13. Http://ontario-wind-resistance.org/member-pages/.

14. Robert Bryce, "Big Wind SLAPPs Critic," *National Review*, June 11, 2013, http://www.nationalreview.com/article/350816/big-wind-slapps-critic-robert-bryce.

15. Jesse Ausubel, "Renewable and Nuclear Heresies," *International Journal of Nuclear Governance, Economy, and Ecology* 1, no. 3 (2007): 233, http://phe.rockefeller.edu/docs/HeresiesFinal.pdf.

16. David J. C. MacKay, *Sustainable Energy—Without the Hot Air* (Cambridge, UK: UIT Cambridge Ltd., 2009), 112. For the full text, go to: http://www.inference.phy.cam.ac.uk/sustainable/book/tex/sewtha.pdf.

17. EIA data, "Levelized Cost of New Generation Sources in the Annual Energy Outlook 2013," January 28, 2013, http://www.eia.gov/forecasts/aeo/electricity_generation.cfm.

18. Vaclav Smil, "Power Density Primer: Understanding the Spatial Dimension of the Unfolding Transition to Renewable Electricity Generation, (Part V–Comparing the Power Densities of Electricity Generation)," Vaclavsmil.com, May 14, 2010, http://www.vaclavsmil.com/wp-content/uploads/docs/smil-article-power-density-primer.pdf, 17.

19. Todd A. Kiefer, "Twenty-First Century Snake Oil: Why the United States Should Reject Biofuels as Part of a Rational National Security Energy Strategy," Waterloo Institute for Complexity and Innovation, January 2013, http://wici.ca/new/wp-content/uploads/2013/02/Kiefer-Snake-Oil2.pdf, 33; 68, note 119.

20. *ScienceDaily*, "Rethinking Wind Power," February 25, 2013, http://www.sciencedaily.com/releases/2013/02/130225121926.html.

21. Amanda S. Adams and David W. Keith, "Are Global Wind Power Resource Estimates Overstated?" *Environmental Research Letters*, February 25, 2013, http://iopscience.iop.org/1748–9326/8/1/015021/pdf/1748–9326_8_1_015021.pdf.

22. Sum for all sixteen projects is 36.605. Divide that sum by 16 = 2.29 W/m².

23. KCET.org, "Explainer: Capacity Factor," undated, http://www.kcet.org/news/rewire/explainers/explainer-capacity-factor.html.

24. EIA data, http://www.eia.gov/electricity/annual/html/epa_04_03.html. In 2011, net summer capacity was 317,640 megawatts.

25. Dieter Helm, *The Carbon Crunch: How We're Getting Climate Change Wrong—and How to Fix It* (New Haven, CT: Yale University Press, 2012), 244.

26. James Hansen, "Baby Lauren and the Kool-Aid," July 29, 2011, http://www.columbia.edu/~jeh1/mailings/2011/20110729_BabyLauren.pdf.

27. Https://plus.google.com/104173268819779064135/posts/Vs6Csiv1xYr.

28. Greenpeace.org, "Stop Climate Change," undated, http://www.greenpeace.org/international/en/campaigns/climate-change/. For details on its budget, see: http://www.greenpeace.org/international/en/about/reports/#a0.

29. Sierraclub.org, "The Goals of the Sierra Club's Climate Recovery Partnership," undated, http://www.sierraclub.org/goals/. See also: John M. Broder, "Sierra Club Leader Will Step Down," *New York Times,* November 18, 2011, http://green.blogs.nytimes.com/2011/11/18/sierra-club-leader-will-step-down/.

30. Christian Azar, Thomas Sterner, and Gernot Wagner, "Rio Isn't All Lost," *New York Times,* June 18, 2012, http://www.nytimes.com/2012/06/19/opinion/rio-isnt-all-lost.html. For EDF's budget, see: http://www.edf.org/finances.

31. PBS Newshour, "Proposed Keystone Pipeline Prompts Protest March, Heated Debate," February 18, 2013, http://www.pbs.org/newshour/bb/environment/jan-june13/keystone_02–18.html. For budget info, see: http://www.nrdc.org/about/finances2012.pdf.

32. Blake Zeff, "Fracking," Branch.com, March 4, 2013, http://branch.com/b/fracking.

33. BP Statistical Review of World Energy, 2013.

34. Ibid.

35. International Energy Agency, "World Energy Outlook 2011," 178. Note that this projection is from the current policies scenario.

36. Http://cleantechnica.com/2011/12/30/german-solar-power-production-surges-60-renewables-20-of-total-electricity-supply/.

37. BP Statistical Review of World Energy 2013.

38. Ibid.

39. American Wind Energy Association data, http://awea.rd.net/Resources/Content.aspx?ItemNumber=5097.

40. BP Statistical Review of World Energy 2013.

41. Germany's land area is almost 350,000 square kilometers. See: http://data.worldbank.org/indicator/AG.LND.TOTL.K2. Note that at 500,000 square kilometers per year, we'd have to cover about 1,000 square kilometers per day. Manhattan Island is about 60 square kilometers. Therefore, just to halt the growth in carbon dioxide emissions with wind energy, we'd have to cover a land area 17 times the size of Manhattan Island, and we'd have to do it *every day.* For the area of Manhattan, see: http://en.Wikipedia.org/wiki/Manhattan.

42. Http://www.upi.com/Science_News/Blog/2013/11/08/Wind-turbines-killed-600000-bats-in-2012/7591383930587/. For birds, see: http://onlinelibrary.wiley.com/doi/10.1002/wsb.260/abstract. This report estimates wind turbines are killing 888,000 bats and 573,000 birds/year (including 83,000 raptor fatalities).

43. T.A. "Ike" Kiefer, "Energy Insecurity," *Strategic Studies Quarterly*, Spring 2013, http://www.au.af.mil/au/ssq/digital/pdf/spring_13/kiefer.pdf, 137.

44. Ibid.

45. This data comes from my last book, *Power Hungry: The Myths of "Green" Energy and the Real Fuels of the Future* (New York: PublicAffairs, 2010), 93.

46. Kiefer, op. cit., 131.

47. The first response, written by Adam L. Rosenberg, whose title is Deputy Director for Technology Strategy, Office of the Assistant Secretary of Defense for Operational Energy Plans and Programs, has just five paragraphs and no footnotes or references to citations. Rosenberg dismissed Kiefer's report as offering "interesting but ultimately misleading opinions." Rosenberg continued, saying the Department of Defense "has a policy of only purchasing operation quantities" of biofuels if they are "cost competitive with conventional fuels." See: http://www.au.af.mil/au/ssq/digital/pdf/spring_13/DOD.pdf.

The other response, from Zia Haq, whose title is Lead Analyst/DPA Coordinator, Department of Energy Bioenergy Technologies Office, has fourteen paragraphs, and not a single substantive citation. Despite this lack of supporting evidence, Haq claimed that Keifer had "tailored" his report by relying exclusively on studies that had "negative points of views and results for biofuels." See: http://www.au.af.mil/au/ssq/digital/pdf/spring_13/DOE.pdf.

48. Nick Taborek, "Pentagon Awards Biofuel Contracts in Obama Renewable Energy Push," *Bloomberg*, May 24, 2013, http://www.bloomberg.com/news/2013-05-24/pentagon-awards-biofuel-contracts-in-obama-renewable-energy-push.html.

49. CNN.com, http://politicalticker.blogs.cnn.com/2011/01/25/obamas-state-of-the-union-remarks/.

50. Ken Silverstein, "Barack Obama Inc.," *Harper's*, November 2006, 40.

51. Http://www.kicktheoilhabit.org/phase1/index.php.

52. Robert Redford, "Redford: Kicking the Oil Habit," CNN.com, May 30, 2006, http://www.cnn.com/2006/US/05/30/redford.oil/index.html.

53. CNN.com, "A Lively Discussion on Rising Oil Prices," May 17, 2006, http://transcripts.cnn.com/TRANSCRIPTS/0605/17/lkl.01.html.

54. Stone Phillips, "A Simple Solution to Pain at the Pump?" NBCnews.com, May 7, 2006, http://www.msnbc.msn.com/id/12676374/#.Tt0sE5hkjGs.

55. Tom Daschle and Vinod Khosla, "Miles per Cob," *New York Times*, May 8, 2006, http://www.nytimes.com/2006/05/08/opinion/08daschle.html?_r=2&ex=1304740800&en=8193adacb1a73c25&ei=5090&partner=rssuserland&emc=rss.

56. Robert Bryce, "Another Failed Energy Loan," *National Review*, December 8, 2011, http://www.nationalreview.com/content/another-failed-energy-loan.

57. David Roberts, "Al Gore, Movie Star, Talks of His Latest Role," *Grist*, May 24, 2006, http://www.msnbc.msn.com/id/12743273/.

58. Roy Roberson, "Former CIA Director Says Farmers Leading Terrorism Fight," *Southeast Farm Press*, March 8, 2007, http://southeastfarmpress.com/grains /030807-farmers-terrorism/.

59. Robert Zubrin and Gal Luft, "Food vs Fuel: A Global Myth," *Chicago Tribune*, May 6, 2008, http://fdd.typepad.com/fdd/2008/05/food-vs-fuel-a.html.

60. Amory Lovins, "Energy Strategy: The Road Not Taken?" *Foreign Affairs*, October 1976, http://www.foreignaffairs.com/articles/26604/amory-b-lovins/energy -strategy-the-road-not-taken, 82.

61. RMI.org, "Blueprint to the New Energy Era," undated, http://www.rmi.org /rmi/ReinventingFireInfographic.

62. This figure is from 2011. See: BP Statistical Review of World Energy 2012.

63. Oak Ridge National Laboratory, "Biofuels from Switchgrass: Greener Energy Pastures," undated, http://bioenergy.ornl.gov/papers/misc/switgrs.html.

64. Texas covers 262,000 square miles; New York, 47,000 square miles; and Ohio 41,000 square miles. See: http://www.worldatlas.com/aatlas/infopage/usabysiz .html.

65. For more, see: http://www.ifpri.org/about/about_menu.asp.

66. Mark W. Rosegrant, "Biofuels and Grain Prices: Impacts and Policy Responses," International Food Policy Research Institute, May 7, 2008, http://www .ifpri.org/pubs/testimony/rosegrant20080507.asp.

67. Donald Mitchell, "A Note on Rising Food Prices," World Bank, April 8, 2008, 1, http://image.guardian.co.uk/sys-files/Environment/documents/2008/07 /10/Biofuels.PDF.

68. Princeton.edu, http://www.princeton.edu/~tsearchi/.

69. Timothy Searchinger, "A Quick Fix to the Food Crisis," *Scientific American*, June 16, 2011, http://www.scientificamerican.com/article.cfm?id=a-quick-fix -to-the-food-crisis.

70. Ibid.

71. Rob Bailey, "The Trouble with Biofuels: Costs and Consequences of Expanding Biofuel Use in the United Kingdom," Chatham House, April 2013, http://www.chathamhouse.org/sites/default/files/public/Research/Energy,%20 Environment%20and%20Development/0413pp_biofuels.pdf, 2.

72. Jean Ziegler, *Betting on Famine: Why the World Still Goes Hungry* (New York: The New Press, 2013), 182.

73. BP Statistical Review of World Energy 2013.

74. Zeigler, op. cit., 180.

75. BP Statistical Review of World Energy 2013.

76. Ziegler, op. cit., 180.

77. California covers about 424,000 square kilometers. See: http://www.enchanted learning.com/usa/states/area.shtml.

78. Economic Research Service, "Major Uses of Land in the United States, 2002," May 2006, http://www.ers.usda.gov/publications/eib-economic-information -bulletin/eib14.aspx#.UWBMF6t4aRk.

79. Ziegler, op. cit., 187–93.

80. Ibid., 203, 204.

81. Robert Bryce, "Food as Fuel," *Slate*, August 1, 2012, http://www.slate.com /articles/news_and_politics/food/2012/07/drought_and_ethanol_how_congress _mandates_and_the_epa_s_new_policy_are_hurting_americans_.html.

Notes to Chapter 21

1. Charles J. Gans, "International Emmys to Honor Al Gore," *USA Today*, November 19, 2007, http://usatoday30.usatoday.com/life/people/2007–11–18-al-gore _N.htm.

2. Http://www.algore.com/about.html.

3. Http://www.ipcc.ch/publications_and_data/ar4/syr/en/spms2.html.

4. Pilita Clark, "Towards a Standstill," *Financial Times*, September 28, 2011.

5. Indeed, climate models may be, in the words of German meteorologist Hans von Storch, "fundamentally wrong." See: *Spiegel Online International*, "Climate Expert von Storch: Why is Global Warming Stagnating?" June 20, 2013, http:// www.spiegel.de/international/world/interview-hans-von-storch-on-problems -with-climate-change-models-a-906721.html.

6. Mona Charen, "Subsidizing Coastal Disasters," *National Review*, June14, 2013, http://www.nationalreview.com/article/351018/subsidizing-coastal-disasters -mona-charen.

7. IEA says that 2.6 billion people are living without clean cooking facilities and 1.3 billion lack access to electricity. See: http://www.iea.org/topics /energypoverty/.

8. BP Statistical Review of World Energy 2013.

9. Ibid.

10. Since the 1970s, the Saudis have been producing about 8.2 million barrels of oil per day. See: BP Statistical Review of World Energy 2012.

11. Vaclav Smil, "Energy at the Crossroads: Notes for a Presentation at the Global Science Forum Conference on Scientific Challenges for Energy Research," Paris, May 17–18, 2006, http://www.oecd.org/dataoecd/52/25/36760950.pdf.

12. BP data from 2011, http://www.bp.com/liveassets/bp_internet/globalbp /globalbp_uk_english/reports_and_publications/statistical_energy_review _2008/STAGING/local_assets/2010_downloads/2030_energy_outlook_booklet .pdf.

13. EIA, International Energy Outlook 2011, http://www.eia.gov/forecasts /ieo/.

14. Associated Press, "Climate Talk Shifts from Curbing CO2 to Adapting," June 15, 2013.

15. *Economist*, "You're Going to Get Wet," June 15, 2013, http://www.economist .com/news/united-states/21579470-americans-are-building-beachfront-homes -even-oceans-rise-youre-going-get-wet.

16. Note that for natural gas, I used the lowest-cost of the various generation types, which in this case was called "advanced." I also used the lowest-cost ex-ample for coal.

17. EIA data: http://www.eia.gov/forecasts/aeo/electricity_generation.cfm.

18. Natural Gas Vehicles for America data, http://www.ngvc.org/about_ngv/.

19. IEA press release, May 24, 2012, http://www.iea.org/newsroomandevents /news/2012/may/name,27216,en.html.

20. Max Luke, "Nuclear and Gas Account for Most Carbon Displacement Since 1950," Breakthrough Institute, September 3, 2013, http://thebreakthrough .org/index.php/programs/energy-and-climate//nuclear-and-gas-account-for -most-carbon-displacement-since-1950/.

21. NGVC.org, http://www.ngvc.org/about_ngv/ngv_environ.html.

22. IEA, Medium-Term Gas Market Report, June 2013, 51.

23. Ibid., 4.

24. IEA data, http://www.iea.org/newsroomandevents/speeches/130611MTGMR Launchpresentation_slides.pdf.

25. US Senate Committee on Environment and Public Works press release, May 24, 2011, http://www.epw.senate.gov/public/index.cfm?FuseAction=PressRoom .PressReleases&ContentRecord_id=23eb85dd-802a-23ad-43f9-da281b2cd287.

26. Here's the math:

$$60,000 \text{ cubic feet of gas} = 60,000,000 \text{ Btu} = 60,000,000 \text{ J}$$
$$60,000,000 \text{ J} / 86,400 \text{ seconds in a day} = 694,444 \text{ W}$$
$$694,444 \times 0.33 = 229,166 \text{ W}$$
If we assume a 2-acre well site (8094 square meters),
$$\text{then } 229,166 \text{ W} / 8094 = 28.3 \text{ W/m}^2.$$

27. Http://en.wikipedia.org/wiki/Energy_density.

28. The IEA puts recoverable global gas resources at 850 trillion cubic meters. See World Energy Outlook 2009, executive summary, 49. A cubic meter contains 35.3 cubic feet.

29. Global gas consumption is about 117 trillion cubic feet per year. See BP Statistical Review of World Energy 2013.

30. Sarah Young and John McGarrity, "Britain Doubles North England Shale Ggas Estimate," Reuters, June 27, 2013, http://uk.reuters.com/article/2013/06/27 /uk-britain-shale-resources-idUKBRE95Q0CD20130627.

31. World Bank data, "Estimated Flared Volumes from Satellite Data, 2007– 2011, http://web.worldbank.org/WBSITE/EXTERNAL/TOPICS

/EXTOGMC/EXTGGFR/0,,contentMDK:22137498~menuPK:3077311~page
PK:64168445~piPK:64168309~theSitePK:578069,00.html#1.

32. Russia is flaring 37.4 billion cubic meters per year, or about 3.6 billion cubic
feet per day that's nearly equal to France's 2011 consumption of 4 Bcf/d. See: BP
Statistical Review of World Energy 2013.

33. Penn Energy, "NIOC Aims to Cut Natural Gas Flaring in Iran," October
7, 2013, http://www.pennenergy.com/articles/pennenergy/2013/10/nioc-aims-to
-cut-natural-gas-flaring-in-iran.html. In 2011, Belgium was consuming 1.6 Bcf/d.

34. BP Statistical Review of World Energy 2013. Vietnam used about 0.8 Bcf/d
in 2011. The US flared 7.1 billion cubic meters in 2011, which works out to almost
0.7 Bcf/d.

35. BP Statistical Review of World Energy 2013.

36. Tamsin Carlisle, "Trying to End Iraq's Wasteful Natural Gas Flaring," *Platts Oil-
gram News,* June 17, 2013, http://blogs.platts.com/2013/06/17/iraq-flare/?sf426329=1.

37. Ryan Salmon and Andrew Logan, "Flaring Up: North Dakota Natural Gas
Flaring More than Doubles in Two Years," Ceres, July 2013, https://www.ceres
.org/resources/reports/flaring-up-north-dakota-natural-gas-flaring-more-than-
doubles-in-two-years, 5.

38. BP Statistical Review of World Energy 2013.

Notes to Chapter 22

1. Http://www.extremetech.com/extreme/147814-the-nuclear-power-vendetta
-or-the-greatest-environmentalist-hypocrisy-of-all-time/3.

2. Estimate from William Horak, chairman of Brookhaven National Laboratory's nuclear science and technology department. For more on Horak, see: http://
www.bnl.gov/newsroom/news.php?a=11183.

3. For generation capacity, see: http://www.entergy-nuclear.com/plant_information
/indian_point.aspx. For percent of electricity, see: http://www.nytimes.com
/2011/03/17/nyregion/17towns.html.

4. According to the NRDC (http://www.nrdc.org/nuclear/indianpoint/files
/NRDC-1336_Indian_Point_FSr8medium.pdf), Indian Point covers 239 acres.
At 4046 square meters per acre, the plant covers 971,246 square meters. Dividing
2.069 billion watts by that area yields 2130 W/m².

5. To get 2.069 billion watts with wind turbines, you'd need 2 billion square
meters, which is 772 square miles. Rhode Island covers 1,045 square miles. See:
RI.gov data, http://sos.ri.gov/library/history/facts/.

6. World Nuclear Association data, http://world-nuclear.org/info/Facts-and
-Figures/Heat-values-of-various-fuels/#.Ud3sIz54ZDs.

7. EIA data, http://www.eia.gov/todayinenergy/detail.cfm?id=5250.

8. EIA data on power plant capital costs, April 2013, http://www.eia.gov/forecasts
/capitalcost/pdf/updated_capcost.pdf, 6.

9. Author interview with project managers from MEAG Power, one of the entities investing in the Vogtle project.

10. DK Eyewitness Books, *Oil: Discover the Story of Petroleum, and the Many Ways It Shapes the World We Live In*, New York: DK Publishing, 2007, 67–68.

11. Http://books.google.com/books?id=dknlH_8XRRwC&pg=PA65&lpg =PA65&dq=poland+oil+lamps+and+carpathian+mountains&source=bl&ots =mp9E7dV6Vo&sig=dYPmf5SIi3alFTYiBBsjFsLSnWQ&hl=en&sa=X&ei=NR _oUa35Aoj7rAHEgYHIDA&ved=oCFQQ6AEwBQ#v=onepage&q=poland%20 oil%20lamps%20and%20carpathian%20mountains&f=false.

12. San Joaquin Valley Geology, "The History of the Oil Industry," undated, http://www.sjvgeology.org/history/index.html.

13. Vaclav Smil, *Energies: An Illustrated Guide to the Biosphere and Civilization* (Cambridge, MA: MIT Press, 1999), 123.

14. SEIA.org, http://www.seia.org/policy/solar-technology/photovoltaic-solar -electric.

15. About.com, http://archaeology.about.com/od/ancientdailylife/qt/fire _control.htm.

16. Pennsylvania Historical & Museum Commission data, http://www.portal. state.pa.us/portal/server.pt/community/history/4569/it_happened_here/471309.

17. Paul Brown, "First Nuclear Power Plant to Close," *Guardian*, March 21, 2003, http://www.guardian.co.uk/uk/2003/mar/21/nuclear.world.

18. Http://en.wikipedia.org/wiki/Shippingport_Reactor.

19. Greenpeace.org, http://www.greenpeace.org/usa/en/campaigns/nuclear/.

20. Sarah Fecht, "One Year Later: A Fukushima Nuclear Disaster Timeline," *Scientific American*, March 8, 2012, http://www.scientificamerican.com/article.cfm ?id=one-year-later-fukushima-nuclear-disaster.

21. Reuters, "Insight: Japan's 'Long War' to Shut Down Fukushima," March 8, 2013, http://www.reuters.com/article/2013/03/08/us-japan-fukushima-idUSBRE 92417Y20130308.

22. Mark Holt, Richard Campbell, and Mary Beth Nikitin, "Fukushima Nuclear Disaster," Congressional Research Service, January 18, 2012, http://www.fas .org/sgp/crs/nuke/R41694.pdf, 1. See also, Beyondnuclear.org, http://www.beyond nuclear.org/home/2011/8/2/two-fukushima-daiichi-workers-drowned-by -tsunami-had-been-or.html.

23. World Health Organization, "Health Risk Assessment from the Nuclear Accident After the 2011 Great East Japan Earthquake and Tsunami," February 2013, http://apps.who.int/iris/bitstream/10665/78218/1/9789241505130_eng .pdf.

24. *The Australian*, "No Fukushima Radiation Problems: Report," June 1, 2013, http://www.theaustralian.com.au/news/latest-news/no-radiation-problems-from -fukushima-rep/story-fn3dxix6-1226654915469.

25. Hiroko Tabuchi, "Nuclear Operator Raises Alarm on Crisis," *New York Times,* August 23, 2013, http://www.nytimes.com/2013/08/24/world/asia/nuclear -operator-raises-alarm-on-crisis.html?_r=0.

26. Reuters, "Insight: Japan's 'Long War' to Shut Down Fukushima," March 8, 2013, http://www.reuters.com/article/2013/03/08/us-japan-fukushima-idUSBRE 92417Y2013030.

27. Miles Doran, "Japan Earthquake: How Big Was It?" CBSnews.com, March 11, 2011, http://www.cbsnews.com/8301–503543_162–20042270–503543/japan-earth quake-how-big-was-it-/.

28. Alan Buis, "Japan Quake May Have Shortened Earth Days, Moved Axis," NASA.gov, March 14, 2011, http://www.nasa.gov/topics/earth/features/japan quake/earth20110314.html.

29. National Police Agency of Japan data, June 9, 2013, http://www.npa.go.jp /archive/keibi/biki/higaijokyo_e.pdf.

30. George Monbiot, "Why Fukushima Made Me Stop Worrying and Love nuclear Power," *Guardian,* March 21, 2011, http://www.guardian.co.uk/comment isfree/2011/mar/21/pro-nuclear-japan-fukushima.

31. Kevin Bullis, "Safer Nuclear Power, at Half the Price," *Technology Review,* March 12, 2013, http://www.technologyreview.com/news/512321/safer-nuclear -power-at-half-the-price/.

32. For more, see http://transatomicpower.com/.

33. Bullis, op. cit.

34. Leslie Dewan from Transatomic Power, speaking at TEDx New England, November 1, 2011, http://www.youtube.com/watch?feature=player_embedded&v =AAFWeIp8JTo#at=779.

35. Wikipedia.org, http://en.wikipedia.org/wiki/Experimental_Breeder_Reactor_II.

36. GE-energy.com, http://www.ge-energy.com/products_and_services/products /nuclear_energy/prism_sodium_cooled_reactor.jsp.

37. Tim Dean, "New Age Nuclear," *Cosmos,* April 2006, http://www.cosmos magazine.com/node/348/full.

38. LightBridge presentation data January 18, 2013, http://files.shareholder. com/downloads/AMDA-16UEEM/2574200371x0x630146/c6267fc1-e0ab-4281 -b8a8-40c3e30df1d6/Lightbridge+-+MNA+Symposium_Public.pdf.

39. GoogleFinance data, https://www.google.com/finance?q=ltbr&ed=us&ei =fpLcUejiNceTqQGg2AE.

40. MarketWatch, "Bill Gates Invests $35 Million in TerraPower's Nuclear Effort," May 25, 2010, http://www.huffingtonpost.com/2010/06/17/bill-gates-invests -35-mil_n_615812.html.

41. Terrapower.com data, http://terrapower.com/people/nathan-myhrvold.

42. Terrapower.com data, http://terrapower.com/pages/traveling-wave-reactor.

43. For more, see: http://robertstoneproductions.com/pandoras-promise/.

44. Jessica Lovering, Ted Nordhaus, and Michael Shellenberger, "Out of the Nuclear Closet: Why It's Time for Environmentalists to Stop Worrying and Love the Atom," *Foreign Policy,* September 7, 2012, http://www.foreignpolicy.com/articles/2012/09/07/out_of_the_nuclear_closet?page=full.

45. Ted Nordhaus, Jessica Lovering, and Michael Shellenberger, "How to Make Nuclear Cheap: Safety, Readiness, Modularity, and Efficiency," Breakthrough Institute, July 9, 2013, http://thebreakthrough.org/index.php/programs/energy-and-climate/how-to-make-nuclear-cheap/, 21.

46. Ibid., 58.

47. The Acheson-Lilienthal Report, March 16, 1946.

48. IAEA data, http://www.iaea.org/About/index.html.

49. Government Accountability Office, "Nuclear Nonproliferation: IAEA Has Made Progress in Implementing Critical Programs but Continues to Face Challenges," May 2013, http://www.gao.gov/assets/660/654714.pdf, 7.

50. USgovernmentspending.com, http://www.usgovernmentspending.com/spending_chart_2001_2018USr_14s1li111mcn_.30t.

51. Richard K. Lester, "Nuclear Governance and Nuclear Innovation After Fukushima," MIT.edu, October 22, 2011, http://web.mit.edu/nse/lester/media/GNES_Talk.pdf.

52. White House, "Remarks by President Barack Obama," Prague, Czech Republic, April 5, 2009, http://www.whitehouse.gov/the_press_office/Remarks-By-President-Barack-Obama-In-Prague-As-Delivered/.

53. In 2000, Iran's electricity use was 119 terawatt-hours. In 2012, it was 251 TWh. See: BP Statistical Review of World Energy 2013.

54. BP Statistical Review of World Energy 2013.

55. Federation of American Scientists, "Status of World Nuclear Forces," undated, http://www.fas.org/programs/ssp/nukes/nuclearweapons/nukestatus.html.

56. Jeff McMahon, "U.S. Runs Out of Nuclear Fuel from Russian Warheads," *Forbes,* June 30, 2013, http://www.forbes.com/sites/jeffmcmahon/2013/06/30/u-s-runs-out-of-nuclear-fuel-from-russian-warheads/.

Notes to Chapter 23

1. Http://www.bizjournals.com/sanjose/blog/2012/09/smallest-electric-vehicle-boosted.html.

2. US Census Bureau data from 2008, see: http://www.census.gov/econ/small bus.html.

3. A poll done in France found that for people under the age of thirty, some 70 percent "yearned for risk-free careers as government officials." See: Pascal Bruckner, *The Fanaticism of the Apocalypse,* (Malden, MA: Polity, 2013), 66.

4. For an excellent essay on this term, see Mike Lofgren, "American Exceptionalism: Alibi of a Nation," TruthOut, June 11, 2013, http://truth-out.org/opinion /item/16878-american-exceptionalism-alibi-of-a-nation. Lofgren says that the line "is simply the cover story and alibi for the extraordinary levels of violence that the United States has been exporting for decades."

5. USdebtclock.org, current as of June 2013.

6. Costofwar.org.

7. Government Accountability Office, "Financial Audit: US Government's Fiscal Years 2012 and 2011 Consolidated Financial Statements," January 13, 2013, http://www.gao.gov/assets/660/651357.pdf.

8. Brad Plumer, "America's Staggering Defense Budget, in Charts," *Washington Post*, January 7, 2013, http://www.washingtonpost.com/blogs/wonkblog /wp/2013/01/07/everything-chuck-hagel-needs-to-know-about-the-defense -budget-in-charts/.

9. Peter Huber, "Digital Innovators vs. the Patent Trolls," Wall Street Journal, April 18, 2011.

10. Roy Strom, "Wi-Fi Case Sheds Light on Patent Trolls," Chicago Lawyer, April 1, 2013, http://chicagolawyermagazine.com/Archives/2013/04/Innovation -Patent-Trolls.aspx.

11. Alex Tabarrok, *Launching the Innovation Renaissance* (New York: TED Conferences, 2011).

12. Stephanie Simon, "US Spends Big on Education, but Results Lag Many Nations: OECD," June 25, 2013, Reuters, http://www.reuters.com/article/2013 /06/25/us-usa-education-oecd-idUSBRE95O0CN20130625.

13. CIA data, https://www.cia.gov/library/publications/the-world-factbook/rank order/2078rank.html.

14. Will Englund, "Teens Choosing Death in Russia," *Washington Post*, March 7, 2012, http://articles.washingtonpost.com/2012–03–07/world/35449709 _1_nastya-suicide-russian-teenagers.

15. Lyudmilla Alexandrova, "Russia's Child, Teenage Suicide Rates Highest in Europe," Indrus.in, March 16, 2013, http://indrus.in/society/2013/03/16/russias _child_teenage_suicide_rates_highest_in_europe_22979.html.

16. Ian Bremmer and Nouriel Roubini, "Ian Bremmer and Nouriel Roubini Unveil the New Abnormal," *Institutional Investor*, June 17, 2013, http://www .institutionalinvestor.com/Article/3218470/Banking-and-Capital-Markets /Ian-Bremmer-and-Nouriel-Roubini-Unveil-the-New-Abnormal.html# .UcDEe_Z4ZDu.

17. Abigail Haworth, "Why Have Young People in Japan Stopped Having Sex?" *Guardian*, October 19, 2013, http://www.theguardian.com/world/2013 /oct/20/young-people-japan-stopped-having-sex.

18. For the latest data, see the US Department of Agriculture's World Agricultural Supply and Demand Estimates, http://www.usda.gov/oce/commodity/wasde/latest.pdf.

19. For more, see: Wikipedia, http://en.wikipedia.org/wiki/International_use_of_the_US_dollar. Regarding Zimbabwe, see: Lydia Polgreen, "Using US Dollars, Zimbabwe Finds a Problem: No Change," New York Times, April 24, 2012, http://www.nytimes.com/2012/04/25/world/africa/using-us-dollars-zimbabwe-finds-a-problem-no-change.html?pagewanted=all&_r=0.

20. National Venture Capital Association data, http://www.nvca.org/index.php?option=com_content&view=article&id=344&Itemid=103.

21. National Science Foundation press release, November 26, 2012, http://www.nsf.gov/news/news_summ.jsp?cntn_id=126135.

22. CIA data, https://www.cia.gov/library/publications/the-world-factbook/rankorder/2001rank.html.

23. Nobelprize.org, http://www.nobelprize.org/nobel_prizes/lists/year/. Cross-referenced with list of winners on Wikipedia.org, http://en.wikipedia.org/wiki/List_of_Nobel_laureates_by_country. Data was current as of June 2013.

24. Ibid.

25. Http://www.nobelprize.org/nobel_prizes/facts/.

26. Wikipedia.org, http://en.wikipedia.org/wiki/List_of_Nobel_laureates_by_country.

27. US Census Bureau data. In 2013, the United States had 316 million people out of a global population of 7 billion.

28. Eurostat.ec, press release, May 27, 2013, http://epp.eurostat.ec.europa.eu/cache/ITY_PUBLIC/8-27052013-AP/EN/8-27052013-AP-EN.PDF.

29. IEA data, Key World Energy Statistics 2012, http://www.iea.org/publications/freepublications/publication/kwes.pdf, 42–3. For more recent data, see EIA data, http://www.eia.gov/electricity/monthly/update/end_use.cfm. In March 2013, the average price of electricity was 9.7 cents, with residential electricity costing 11.6 cents, commercial costing 10 cents, and industrial costing 6.7 cents.

30. *Financial Times*, "US Ahead of Europe on Energy Policy," May 13, 2012, http://www.ft.com/intl/cms/s/0/45afd57a-9abf-11e1-9c98-00144feabdc0.html#axzz1urNAcPVV.

31. IEA Key World Energy Statistics 2013, 11–27.

32. EIA data, http://www.eia.gov/dnav/pet/pet_move_exp_dc_NUS-Z00_mbblpd_m.html. In mid-2013, the United States was exporting 3.9 million barrels of oil per day, mostly in the form of diesel fuel.

33. IEA Key World Energy Statistics 2013, op. cit.

34. BP Statistical Review of World Energy 2012. In 2011, China produced about 2,427 million tons of oil equivalent of energy, including 1,956 Mtoe in coal, 92 Mtoe in gas, 203 Mtoe in oil, 19.5 Mtoe in nuclear, and 157 Mtoe from hydro.

35. BP Statistical Review of World Energy 2013.

36. Congressional Research Service, "US Fossil Fuel Resources: Terminology, Reporting, and Summary," March 25, 2011, 14, http://assets.opencrs.com/rpts /R40872_20110325.pdf.

37. OPEC data, http://www.opec.org/opec_web/en/data_graphs/330.html.

38. EIA data, http://www.eia.gov/todayinenergy/detail.cfm?id=10711&src=email.

39. See, for instance, BP Statistical Review 2013 slidepack, http://www.bp.com/content/dam/bp/ppt/Statistical-Review-2013/statistical_review_of_world_energy _2013.pptx, 19.

40. Debi Durham, "Iowa Defends Orascom Fertilizer Plant Incentives," *Quad City Times*, February 28, 2013, http://qctimes.com/news/opinion/editorial /columnists/iowa-defends-orascom-fertilizer-plant-incentives/article_f5ac 78ba-815b-11e2-a9f1-001a4bcf887a.html. For the project's total value, see Ventures -africa.com, August 29, 2013, http://www.ventures-africa.com/2013/08/work -orascoms-iowa-fertiliser-plant-schedule/.

41. *The Business Journal*, "Vallourec Execs Hail 'Great Day' for Youngstown Mill," June 12, 2013, http://businessjournaldaily.com/company-news/vallourec -execs-hail-great-day-youngstown-mill-2013-6-12.

42. BP Statistical Review of World Energy 2013. Russia came in second at about 57.1 billion cubic feet of gas per day.

43. BP Statistical Review of World Energy 2013.

44. EIA data, Short-term energy outlook, February 14, 2013, http://www.eia .gov/forecasts/steo/special/pdf/2013_sp_02.pdf?scr=email. Note that some estimates, including the BP Statistical Review 2013, use a higher number. BP put the increase at 1 million barrels per day.

45. Geology.com, http://geology.com/rocks/shale.shtml.

46. EIA, "Technically Recoverable Shale Oil and Shale Gas Resources: An Assessment of 137 Shale Formations in 41 Countries Outside the United States," June 10, 2013, http://www.eia.gov/analysis/studies/worldshalegas/.

47. WTRG.com, http://www.wtrg.com/rotaryrigs.html. See also Baker Hughes data, http://phx.corporate-ir.net/phoenix.zhtml?c=79687&p=irol -rigcountsoverview.

48. Pipeline101.com, http://www.pipeline101.com/overview/natgas-pl.html.

Notes to Chapter 24

1. Michael Tobias, "The Ehrlich Factor: A Brief History of the Fate of Humanity, with Dr. Paul R. Ehrlich," *Forbes,* January 16, 2013, http://www.forbes.com /sites/michaeltobias/2013/01/16/the-ehrlich-factor-a-brief-history-of-the-fate -of-humanity-with-dr-paul-r-ehrlich/.

2. Christine Shearer, "The Climate Movement Takes on Fracking: Interview With Bill McKibben," Truthout, April 17, 2012, http://truth-out.org/opinion

/item/8518-the-climate-movement-takes-on-fracking-interview-with-bill
-mckibben.

3. Michael Brune, "The Opposition's Opening Remarks," *The Economist*, February 5, 2013, http://www.economist.com/debate/days/view/934.

4. Associated Press, "'March Against Monsanto,' Protesters Rally Against U.S. Seed Giant and GMO Products," May 25, 2013, http://www.huffingtonpost.com /2013/05/25/march-against-monsanto-gmo-protest_n_3336627.html.

5. YouTube, "Why March Against Monsanto?" May 19, 2013, http://www .youtube.com/watch?v=vnKtv2GDEyA.

6. *Fortune*, "World's Most Admired Companies 2013," undated, http://money.cnn .com/magazines/fortune/most-admired/2013/snapshots/670.html. For market capitalization information, see Google Finance. Data current as of July 2013.

7. Michael Shellenberger and Ted Nordhaus, "Evolve: A Case for odernization as the Road to Salvation," *Orion*, September/October 2011, http://www.orion magazine.org/index.php/articles/article/6402/.

8. Mark Lynas, *The God Species: Saving the Planet in the Age of Humans* (Washington, DC: National Geographic, 2011), 243–244.

Notes to Appendix C

1. This assumes an average male can sustain power output of 100 watts. That may be generous. An elite cyclist can sustain nearly 400 watts over the course of several hours. See: http://mapawatt.com/2009/07/19/bicycle-power-watts/. It also assumes a male weight of 170 pounds, which was Lance Armstrong's weight when he was champion. It's also the approximate average weight of the average resident of North America. See: http://voices.yahoo.com/fun-facts-2009-tour -de-france-3700950.html.

2. Answers.com, which puts average at 900 to 1100 lbs. A 1000 lb horse would weigh 454 kilos. This calculation assumes 450 kg.

3. C. Morris, *The Dawn of Innovation: The First American Industrial Revolution* (New York: PublicAffairs, 2012), 110.

4. Ibid., 177.

5. Ibid., 102.

6. John H. White, *A History of the American Locomotive: Its Development, 1830–1880* (New York: Dover Publications, 1980), 75–6. See: http://books.google.com /books?id=1A4iiGAz628C&pg=PA497&lpg=PA497&dq=Cleveland, +Columbus+%26+Cincinnati+Railroad+engine+specs&source=bl&ots =ps2xxrsZOr&sig=UpDbFFRPyUTEbXoDS1ixZUAmuug&hl=en&sa=X &ei=2kPjUM_pHoXu2QXE20C4Cw&ved=0CD4Q6AEwBA#v=one page&q=horsepower&f=false.

7. C. Morris, *The Dawn of Innovation*, 232. Note the calculation only counts the weight of the flywheel. The weight of the engine itself was not available.

For more, see: http://twain.lib.virginia.edu/yankee/cymach1.html. For displacement, see: http://www.newsm.org/steam-engines/corliss-centenial.html. Note that stroke is 10 feet (304.8 cm) while diameter is 44 inches (112 cm). Therefore, displacement—pi × 305 cm × (56 cm)2 = 3M cm3, or 3000 l for each cylinder. Total displacement was 6,000 l.

8. Vaclav Smil, *Prime Movers of Globalization: The History and Impact of Diesel Engines and Gas Turbines* (Cambridge, MA: MIT Press, 2010), 29–30.

9. Http://en.wikipedia.org/wiki/Wright_Flyer#Specifications_.28Wright _Flyer.29; for displacement, see: http://www.wright-brothers.org/Information _Desk/Just_the_Facts/Engines_&_Props/1903_Engine.htm.

10. Http://www.modelt.ca/faq-fs.html; power output from T. Collings, *The Legendary Model T Ford: The Ultimate History of America's First Great Automobile* (Iola, WI: Krause Publications, 2007), 136. See: http://books.google.com /books?id=r61b2-M1zr4C&pg=PA136&lpg=PA136&dq=model+t+engine+and+ 22+hp&source=bl&ots=RiN3nOgDfT&sig=RwB3hypsbMxKnfJ4asDlIc MU9ZA&hl=en&sa=X&ei=zSGzUuaNNIOQ2gXir4GAAg&ved=0CLABEO gBMA4#v=onepage&q=model%20t%20engine%20and%2022%20hp&f =false.

11. Http://en.wikipedia.org/wiki/Wright_R-3350_Duplex-Cyclone.

12. Http://www.carmagazine.co.uk/Drives/Search-Results/First-drives/Ford -Focus-10-Ecoboost-review/.

13. F1technical.net, "Formula One Engines," undated, http://www.f1technical .net/articles/4; Formula1.com, "Formula One Fuel and Oil—F1's Top Secret Tuning Aids," undated, http://www.formula1.com/news/features/2011/9/12524.html.

14. Note that the conversion from thrust to horsepower depends on several factors. This calculation is done using peak speed at sea level. Calculations done by Rex Rivolo, a former military pilot, who holds a PhD in astrophysics. For GEnx data, see: en.wikipedia.org/wiki/General_Electric/Genx.

Notes to Appendix E

1. General Purpose Standing Committee, "Rural Wind Farms," December 2009, http://www.parliament.nsw.gov.au/Prod/parlment/committee.nsf/0/ea24765 9081d31fdca25768e001a2e2a/$FILE/091216%20Report%20-%20Rural%20wind %20farms.pdf, xv.

2. Ibid, xvii.

3. Bruce Rapley and Huub Bakker, eds. *Sound, Noise, Flicker and the Human Perception of Wind Farm Activity* (Palmerston North, New Zealand: Atkinson & Rapley Consulting Ltd, 2010), 7.

4. Http://www.epaw.org/echoes.php?lang=en&article=n19.

5. Http://www.abc.net.au/news/2011–06–27/wind-farm-setbacks-policy-to -remain/2772716.

6. Robert Bryce, "The Party's Over For Big Wind," *Energy Tribune,* August 15, 2011, http://www.energytribune.com/8549/the-partys-over-for-big-wind#sthash.5XnneCAw.dpbs.

7. Lou Wilin, "Can Turbines Generate Health Problems?" *The Courier*, October 5, 2011, http://www.thecourier.com/Issues/2011/Oct/05/ar_news_100511_story4.asp.

8. Http://www.suncor.com/en/about/3680.aspx.

9. Http://www.ert.gov.on.ca/files/201107/00000300-AKT5757C7CO026-BGI54 EDI9RO026.pdf.

10. Http://www.library.ucla.edu/pdf/Chen.Paper.pdf.

11. Michael A Nissenbaum, Jeffery J. Aramini, and Christopher D. Hanning, "Effects of Industrial Wind Turbine Noise on Sleep and Health," *Noise & Health* 2012; 14:237–43, http://www.noiseandhealth.org/text.asp?2012/14/60/237/102961.

Notes to Appendix F

1. Reuters, "E.ON Completes World's Largest Wind Farm in Texas," October 1, 2009, http://www.reuters.com/article/2009/10/01/wind-texas-idUSN 3023624320091001.

2. Graham Lloyd, "What You Can't Hear Can Hurt You," *The Australian*, January 25, 2012, http://www.theaustralian.com.au/news/features/what-you-cant -hear-can-hurt-you/story-e6frg6z6–1226252801681.

3. Nrgenergy.com, http://www.nrgenergy.com/econrg/wind.html.

4. Steve Clark, "Duke Energy Announces Wind Farm for Willacy County," *Brownsville Herald*, August 12, 2011, http://www.brownsvilleherald.com/articles /wind-129933-energy-farm.html.

5. Http://www.semprausgp.com/energy-solutions/wind-flat-ridge2.html.

6. BP.com, http://www.bp.com/liveassets/bp_internet/alternative_energy /alternative_energy_english_new/STAGING/local_assets/downloads_pdfs /Flat_Ridge_1_Wind_Farm_Fact_Sheet110811.pdf.

7. Bureau of Land Management data, http://www.blm.gov/pgdata/etc/medialib /blm/wy/information/NEPA/rfodocs/chokecherry/feis.Par.91515.File.dat/CCSM _Vol_ll-ExSum.pdf.

8. Infigen Energy, "The Capital Wind Farm," undated, http://www.infigen energy.com/Media/docs/Capital-Wind-Farm-Brochure-fb268c8e-e14b-4c44 –8d95-e8ec2a23e31a-0.pdf, 4.

9. Trustpower.co.nz, undated, http://www.trustpower.co.nz/index.php?section =163.

10. Ontario Power Authority data, http://www.powerauthority.on.ca/wind-power /ripley-wind-power-project-76-mw-huronkinloss-twp.

11. Ontario Power Authority data, http://www.powerauthority.on.ca /wind-power/erie-shores-wind-farm-99-mw-port-burwell.

12. Ontario Power Authority data, http://www.powerauthority.on.ca/wind-power /greenwich-windfarm-989-mw-district-thunder-bay.

13. Ontario Power Authority data, http://www.powerauthority.on.ca/wind-power /kingsbridge-i-wind-power-project-396-mw-goderich.

14. Ontario Power Authority data, http://www.powerauthority.on.ca/wind -power/melancthon-i-wind-plant-675-mw-melancthon-twp.

15. BP.com, http://www.bp.com/liveassets/bp_internet/alternative_energy /alternative_energy_english_new/STAGING/local_assets/downloads_pdfs/me hoopany_wind_farm_factsheet.pdf.

SELECT BIBLIOGRAPHY

Abbey, Edward. *Beyond the Wall: Essays from the Outside*. New York: Henry Holt, 1971.

Bejan, Adrian, and Zane, J. Peder. *Design in Nature: How the Constructal Law Governs Evolution in Biology, Physics, Technology, and Social Organization*. New York: Doubleday, 2012.

Brand, Stewart. *Whole Earth Discipline: An Ecopragmatist Manifesto*. New York: Viking, 2009.

Brookes, Tim. *The Guitar: An American Life*. New York: Grove Press, 2005.

Bruckner, Pascal. *The Fanaticism of the Apocalypse*. Malden, MA: Polity, 2013.

Burrell, Brian. *Merriam Webster's Guide to Everyday Math: A Home and Business Reference*. Springfield, MA: Merriam Webster, 1998.

Cardwell, Donald. *Wheels, Clocks, and Rockets: A History of Technology*. New York: W.W. Norton, 1995.

Chernow, Ron. *Titan: The Life of John D. Rockefeller, Sr*. New York: Vintage Books, 1998.

Colwell, Robert P. *The Pentium Chronicles: The People, Passion, and Politics Behind Intel's Landmark Chips*. Hoboken, NJ: John Wiley & Sons, 2006.

Cravens, Gwyneth. *Power to Save the World: The Truth About Nuclear Energy*. New York: Knopf, 2008.

Dyson, George. *Turing's Cathedral: The Origins of the Digital Universe*. New York: Pantheon, 2012.

Editors of *Bicycling* magazine, *The Noblest Invention: An Illustrated History of the Bicycle*. United States: Rodale, 2003.

Evans, Harold. *They Made America: From the Steam Engine to the Search Engine: Two Centuries of Innovators*. New York: Little, Brown and Company, 2004.

Freese, Barbara. *Coal: A Human History*. New York: Penguin, 2003.

Freund, Paul, and Olav Kaarstad. *Keeping The Lights On: Fossil Fuels in the Century of Climate Change*. Oslo: Universitetsflorlaget, 2007.

Glaeser, Edward. *Triumph of the City: How Our Greatest Invention Makes Us Richer, Smarter, Greener, Healthier, and Happier*. New York: Penguin, 2011.

Goodell, Jeff. *Big Coal: The Dirty Secret Behind America's Energy Future*. Boston: Houghton Mifflin, 2006.

Hamilton, Tyler, and Daniel Coyle. *The Secret Race: Inside the Hidden World of the Tour de France: Doping, Cover-ups, and Winning at All Costs*. New York: Bantam Books, 2012.

Hayden, Howard C. *The Solar Fraud: Why Solar Energy Won't Run The World*. Pueblo West, CO: Vales Lake Publishing LLC, 2001.

Helm, Dieter. *The Carbon Crunch: How We're Getting Climate Change Wrong—and How to Fix It*. New Haven, CT: Yale University Press, 2012.

Hill, Donald, *A History of Engineering in Classical and Medieval Times*. London: Routledge, 1984.

Huber, Peter W., and Mark P. Mills. *The Bottomless Well: The Twilight of Fuel, the Virtue of Waste, and Why We Will Never Run Out of Energy*. New York: Basic Books, 2005.

Jevons, William Stanley. *The Coal Question: An Inquiry Concerning the Progress of the Nation, and the Probable Exhaustion of our Coal Mines*. Dodo Press. 1866.

Johnson, Steven. *Where Good Ideas Come From: The Natural History of Innovation*. New York: Riverhead, 2010.

Klein, Maury. *The Power Makers: Steam, Electricity, and the Men Who Invented Modern America*. New York: Bloomsbury, 2008.

Kotkin, Joel. *The City: A Global History*, New York: The Modern Library, 2005.

Landes, David S. *Dynasties: Fortunes & Misfortunes of the World's Great Family Businesses*. New York: Penguin, 2006.

Lovins, Amory B., E. Kyle Datta, Odd-Even Bustnes, Jonathan G. Koomey, and Nathan J. Glasgow. *Winning the Oil Endgame: Innovation for Profits, Jobs, and Security*. Boulder, CO: Rocky Mountain Institute, 2004.

MacKay, David J. C. *Sustainable Energy—Without the Hot Air*. Cambridge: UIT Cambridge Ltd., 2009. Available free online at: http://www.inference.phy.cam.ac.uk/sustainable/book/tex/cft.pdf.

Mackey, John, and Raj Sisodia. *Conscious Capitalism: Liberating the Heroic Spirit of Business*. Boston: Harvard Business Review Press, 2013.

McClellan, James E. III, and Harold Dorn, *Science and Technology in World History: An Introduction*. Baltimore: Johns Hopkins University Press, 2006.

McCullough, David. *The Path Between the Seas: The Creation of the Panama Canal 1870–1914*. New York: Simon & Schuster, 1977.

McKibben, Bill. *Eaarth: Making a Life on a Tough New Planet*. Toronto: Vintage, 2010.

Medina, John. *Brain Rules: 12 Principles for Surviving and Thriving at Work, Home, and School*. Seattle: Pear Press, 2008.

Morozov, Evgeny. *To Save Everything Click Here: The Folly of Technological Solutionism.* New York: PublicAffairs, 2013.

Morris, Charles R. *The Dawn of Innovation: The First American Industrial Revolution.* New York: PublicAffairs, 2012.

Muller, Richard A. *Physics For Future Presidents: The Science Behind the Headlines.* New York: W.W. Norton, 2008.

Nye, David E. *Electrifying America: Social Meanings of a New Technology.* Cambridge MA: MIT Press, 1992.

Owen, David. *The Conundrum: How Scientific Innovation, Increased Efficiency, and Good Intentions Can Make Our Energy and Climate Problems Worse.* New York: Riverhead, 2011.

Pinker, Steven. *The Better Angels of Our Nature: Why Violence Has Declined.* New York: Viking, 2011.

Rapley, Bruce, and Huub Bakker, eds. *Sound, Noise, Flicker and the Human Perception of Wind Farm Activity.* Palmerston North: Atkinson & Rapley Consulting Ltd, 2010.

Ridley, Matt. *The Rational Optimist: How Prosperity Evolves.* New York: HarperCollins, 2010.

Rodale Inc. *The Noblest Invention: An Illustrated History of the Bicycle.* USA: Rodale, 2003.

Rosen, William. *The Most Powerful Idea in the World: A Story of Steam, Industry, and Invention.* New York: Random House, 2010.

Smil, Vaclav. *Creating the Twentieth Century: Technical Innovations of 1867–1914 and Their Lasting Impact.* Oxford: Oxford University Press, 2005.

———. *Energies: An Illustrated Guide to the Biosphere and Civilization.* Cambridge, MA: MIT Press, 1999.

———. *Energy at the Crossroads: Global Perspectives and Uncertainties.* Cambridge, MA: MIT Press, 2003.

———. *Energy Myths and Realities: Bringing Science to the Energy Policy Debate.* Washington, DC: AEI Press, 2010.

———. *Global Catastrophes and Trends: The Next Fifty Years.* Cambridge, MA: MIT Press, 2008.

———. *Oil.* Oxford: Oneworld Publications, 2008.

———. *Prime Movers of Globalization: The History and Impact of Diesel Engines and Gas Turbines.* Cambridge, MA: Massachusetts Institute of Technology, 2010.

Till, Charles E., and Yoon Il Chang. *Plentiful Energy: The Story of the Integral Fast Reactor.* Amazon: CreateSpace, 2011.

Topol, Eric. *The Creative Destruction of Medicine: How the Digital Revolution Will Create Better Health Care.* New York: Basic Books, 2012.

Tucker, William. *Terrestrial Energy: How Nuclear Power Will Lead the Green Revolution and End America's Energy Odyssey.* Savage, MD: Bartleby Press, 2008.

Usher, Abbott Payson. *A History of Mechanical Inventions.* New York: Dover Publications, 1982.

Witherell, James L. *Bicycle History: A Chronological History of People, Races, and Technology.* Cherokee Village, AR: McGann, 2010.

Wolman, David. *The End of Money: Counterfeiters, Preachers, Techies, Dreamers—And the Coming Cashless Society.* Boston: Da Capo Press, 2012.

Ziegler, Jean. *Betting on Famine: Why the World Still Goes Hungry.* New York: The New Press, 2013.

INDEX

Thomas Cooper

Robert Bryce is the acclaimed author of four previous books, including, most recently, *Power Hungry: The Myths of "Green" Energy and the Real Fuels of the Future*. A senior fellow at the Manhattan Institute, his articles have appeared in dozens of publications, including the *Wall Street Journal, New York Times, Washington Post, Guardian, Austin Chronicle, Bloomberg View, Counterpunch,* and *National Review*. An apiarist, he lives in Austin with his wife, Lorin, and their children, Mary, Michael, and Jacob.

PublicAffairs is a publishing house founded in 1997. It is a tribute to the standards, values, and flair of three persons who have served as mentors to countless reporters, writers, editors, and book people of all kinds, including me.

I. F. STONE, proprietor of *I. F. Stone's Weekly*, combined a commitment to the First Amendment with entrepreneurial zeal and reporting skill and became one of the great independent journalists in American history. At the age of eighty, Izzy published *The Trial of Socrates*, which was a national bestseller. He wrote the book after he taught himself ancient Greek.

BENJAMIN C. BRADLEE was for nearly thirty years the charismatic editorial leader of *The Washington Post*. It was Ben who gave the *Post* the range and courage to pursue such historic issues as Watergate. He supported his reporters with a tenacity that made them fearless and it is no accident that so many became authors of influential, best-selling books.

ROBERT L. BERNSTEIN, the chief executive of Random House for more than a quarter century, guided one of the nation's premier publishing houses. Bob was personally responsible for many books of political dissent and argument that challenged tyranny around the globe. He is also the founder and longtime chair of Human Rights Watch, one of the most respected human rights organizations in the world.

·   ·   ·

For fifty years, the banner of Public Affairs Press was carried by its owner Morris B. Schnapper, who published Gandhi, Nasser, Toynbee, Truman, and about 1,500 other authors. In 1983, Schnapper was described by *The Washington Post* as "a redoubtable gadfly." His legacy will endure in the books to come.

Peter Osnos, *Founder and Editor-at-Large*